Nonhuman Primates in Biomedical Research

THE
WESLEY W. SPINK
LECTURES ON
COMPARATIVE MEDICINE
Volume 3

NONHUMAN
PRIMATES
IN
BIOMEDICAL
RESEARCH

WILLIAM MONTAGNA

Foreword by
S. L. WASHBURN

UNIVERSITY OF MINNESOTA PRESS
Minneapolis

Library of Congress Catalog Card Number 76-7881
ISBN 0-8166-0793-1

The Wesley W. Spink Lectures on Comparative Medicine, established in honor of Dr. Spink's wide range of accomplishments, are presented by international authorities in comparative medicine and biology. A graduate of Carleton College and Harvard Medical School, Dr. Spink has maintained a deep interest in comparative medicine for almost forty years at the University of Minnesota, where he is now Regents' Professor Emeritus of Medicine and Comparative Medicine. Sponsorship of the lectures jointly by Carleton College and the University of Minnesota reflects the concern of both institutions for the dissemination of scientific knowledge. The lectures, and the publication of the volumes based on them, have been assisted by a grant from the Bush Foundation.

CONTENTS

FOREWORD

Our closest living relatives, the primates, are a remarkably diversified order of mammals. Understanding this diversity is both a problem and an opportunity. It is a problem because it is very hard to make generalizations about such different forms. It is an opportunity because the variety gives insights and poses fascinating problems. The uniqueness of the situation may be shown by comparison with other groups of mammals. The contemporary horses are all very much alike. Study of the living horses would give no information on the tiny forms of 50 million years ago. Changes in the skull and teeth, number of toes, geographical distribution, and possible behaviors would be known only from the fossil record. Study of the elephants would permit only the most general guesses about their evolution, and only the fossils would show the complex history and forms with four tusks, tusks in the lower jaw, flat tusks, tusks curving back like giant hoes.

Some of the primates have changed very little from the kinds which lived 50 million years ago. Our understanding may be based both on the fossil record and on the biology

and behavior of the present forms. It is as if we could study the behaviors of the earliest horses in the field and then bring them into the laboratory! It is as if we could see a saber-toothed tiger attacking its prey and learn precisely how those mighty teeth were used! How little we would know of strange primates like the aye-aye or *Tarsius* if the evidence were limited to fossils. But when the tiny tarsier becomes active in the evening and leaps to catch an insect, we see the meaning of its huge eyes, long legs, and primitive dentition.

History divided the primates so that there are three great natural experiments as well as innumerable smaller ones. Many of the most primitive primates were isolated on the island of Madagascar. There they evolved, free from the fierce competition of the African mainland. In science fiction one can ask what would have happened to the earliest horses had they not been subjected to strong selection. But the answer would only be fiction. The diversity of the lemurs on Madagascar provides an actual case of what really did happen to protected early primates. In a similar way the South American monkeys evolved independently of the Old World monkeys for many millions of years, and they too offer a set of experiments in monkey evolution, greatly enriching the understanding we would have if information were limited to the Old World alone.

From a theoretical point of view, we are exceedingly fortunate that our nearest relatives are so highly diversified. They offer fascinating information on evolution, behavior, locomotion, reproduction, and disease — to mention only a few of the many problems facing science. From a medical point of view, this diversity offers many opportunities which would not exist if primate evolution had resulted in the survival of only a few very similar forms. From an immediate point of view, we are fortunate that these lectures were given by Wil-

liam Montagna, who has wisely selected a few important topics from the vast body of primate information and has elected to stress the importance of the primates in medical research. His own research on the skin has led him to study all the major kinds of living primates, giving him a background very few other scientists can match.

While stressing the practical and intellectual utility of the primates to man, Montagna also clearly recognizes man's obligation to protect and conserve the diversity of these animals. For the first time in the history of life, a history that stretches back more than 3 billion years, a single species can now determine the fate of countless others. As Montagna points out, as human populations multiply and farming and civilizations spread, the forests which support the primates are inevitably reduced. Even now, many primates are in danger of extinction, and there is every indication that the situation will become much worse in the near future. One hopes these lectures will interest people in the primates and their importance in medical and behavioral research. One also hopes they will alert people to the need for international cooperation in protecting the diversity of primates, in protecting the irreplaceable heritage of 70 million years of evolution.

S. L. Washburn

University of California, Berkeley
March 1976

ACKNOWLEDGMENTS

I extend my deepest thanks to Dr. Wesley W. Spink and the sponsors of the Lectures on Comparative Medicine for the honor they have bestowed upon me in asking me to give the third lectures of the series. These lectures are the basis of this book.

I acknowledge with gratitude the thoughtful help I received from many of my colleagues at the Oregon Regional Primate Research Center, principally from Dr. Theodore Grand, Dr. Wilbur P. McNulty, Jr., Dr. Oscar W. Portman, Dr. M. R. Malinow, Dr. Charles H. Phoenix, and Dr. Charles Howard; and from Dr. Leon H. Schmidt of the Kettering-Meyer Laboratory, Southern Research Institute, Birmingham, Alabama. My very special thanks to Mrs. Margaret Barss for her continued unstinting devotion to our work and to Mrs. Margaret Shininger for her selfless help throughout the preparation of these lectures.

William Montagna

Oregon Regional Primate Research Center, Beaverton
March 1976

Nonhuman Primates in Biomedical Research

INTRODUCTION

T he veritable floods of primatological observations that daily inundate and threaten to overwhelm us make it more difficult to write briefly about the Order Primates than it was even a few decades ago when most of the data lay buried in obscure tomes, many of them written largely in the German language. Today, numerous publications appear at frequent intervals in journals, monographs, and book series which are dedicated entirely to primate biology and to promoting the use of nonhuman primates in medical research. So much has already been written about the biology of primates that the major problem confronting any author is to decide what to include and what to eliminate. In addition, the Order Primates is so large and diversified that unless one envisions writing a book of encyclopedic proportions, he must resist the temptation to be all-inclusive and must judiciously limit himself to selected subjects. For this reason, the following chapters are intentionally brief and highly selective. Reflecting my own preoccupations, they are intended to whet the appetite rather than exhaust the reader.

3

Now that the number of nonhuman primates available for investigation is rapidly and perilously diminishing, we are confronted with the difficult problem of balancing demand and supply. Biomedical investigators who use primates for experimentation would be well-advised to plan now to ensure a supply for the future, or they may be using the last of these valuable animals.

Just the word *primates* conjures up to some biomedical investigators the image of a beast whose looks and behavior somewhat resemble those of man and who, therefore, is a suitable model for experimentation. Yet not all nonhuman primates are necessarily good subjects in health-oriented research. For example, whereas studies of the gastrointestinal physiology of Old or New World leaf-eating monkeys might very well add to the bulk of biological lore, they would not ipso facto engender much light on the subject of human gastric physiology. For all practical purposes, these animals, especially the Old World Colobinae, are ruminants. Langurs or Old World leaf-eaters, for example, have a pouched stomach (Figure 1), only one of the many striking biological disparities that characterize the members of this order. Like the bats, all of which are recognizable as bats and behave somewhat alike, primates have highly diversified structural, physiological, and behavioral properties. The liver of a leaf-eating monkey is equipped with enzymes which must be different from those of macaques, just as the digestive physiology of a vampire is remarkably different from that of a fish-eating bat or a flying fox. Notwithstanding these diversities, however, one factor unifies all nonhuman primates, including even the prosimians: New World or Old; arboreal or terrestrial; agile or lethargic; herbivorous, carnivorous, or omnivorous, they all have an extremely large brain in terms of mass per body weight. It is

Fig. 1. The pouched stomach of a langur grossly resembles
that of a ruminant. (Courtesy of Dr. Walter S. Tyler, director,
University of California Primate Research Center,
Davis, California.)

this characteristic which allies them so closely to man, more
closely, in fact, than to any other mammal.

In those areas of medical research where ethical and moral
considerations preclude the use of volunteer human beings as
experimental subjects, nonhuman primates are often the only
feasible substitute. Some, perhaps even many, investigators
are aware of this fact, but its immense potential for improving human health has scarcely begun to be exploited. Perhaps

the failure to do so can be traced partly to the fact that primate husbandry is a difficult art and partly to the realization that to extend the use of these animals in medical research may contribute to their extinction.

These chapters are intended to introduce the investigator to the biological and social properties of primates and to show how the catastrophe just mentioned can be averted. I make no apologies for the many omissions but hope that the interested scholar will be motivated enough by what he reads here to explore on his own whatever is of interest. Like granaries in the fall, the research libraries of the world are pressed down and overflowing with a wealth of information, and field research is almost as accessible.

CHAPTER 1

ANIMALS FOR
BIOMEDICAL RESEARCH

S ince the beginning of time, man has believed in the per-
fectibility of his biological as well as of his spiritual being.
And as the eons have rolled by, that faith in himself has
seemed to be well founded since man is basically a tough animal. Yet, in our more sober moments, we can see that all is
not well with man. Whether or not we believe in a divine cre-
ator who fashioned us after His own image, we have to con-
cede that we are animals, thinking animals, and as such have
had to accept the liabilities as well as the assets, the agonies
as well as the ecstasies of our human condition. Man's striving
to attain in actuality the perfection he deems himself capable
of is forever thwarted by a deep awareness of his imperfec-
tion. Thus, born to reach for the stars, he is pulled earthward
by his own weight. Reflect, for example, that each of our
unique assets—upright posture and large brain, consciousness,
intelligence, curiosity, and sense of adventure; a long gestation, prolonged immaturity, extraordinary mobility and en-

NOTE: This chapter is based on a lecture in the Wesley W. Spink Lectures on
Comparative Medicine presented at Carleton College, Northfield, Minnesota, on
October 13, 1975.

durance, and great longevity; and finally our astonishingly complex social order and means of communication — is also a liability. Man pays dearly for being man. In addition to his naturally occurring defects and biological blunders, he has added to his own problems by overtaxing his physical and mental attributes to the breaking point and by ensuring the survival and reproduction of individuals whose genetic constitutions are so faulty that the whole console of man's ingenuity is called into play to preserve them in existence.

Thus, driven by an inner compulsion to exceed his grasp, yet equipped with a naturally defective carcass and a zeal to correct its faults, man has applied his genius to strengthening his physical weaknesses and even to altering his biological attributes to his own advantage. His unending struggle with the debilities that attend old age and his futile attempts to stave off death are but two examples. Certainly, man has progressively increased his longevity, but to what purpose? True, his modest success in prolonging life has given to millions of elderly the hope that their demise can be delayed indefinitely. But in the meantime most of them suffer cruelly and needlessly during the last vegetative years of a long, too often impoverished old age.

Medical research exists because we are fundamentally constructed badly, because we are an easy prey to invading organisms and the victim of our own or someone else's foolishness and carelessness, and because our inhumanity to one another makes us constant predators of our own species. During most of our normally long lives, we are intent on preventing damage to ourselves and others and on repairing damage once inflicted. Consequently, medicine has become an integral part of our culture and is the direct result of our ability to analyze, record, store, and retrieve past experiences.

The most direct, therefore the most obvious, approach to

our medical problems is the observation of man himself. But there are severe limitations to the insults which we can inflict on our fellow human beings in our search for the answers to these problems. Hence, in the many areas of medical research where ethical and moral considerations preclude the use of human beings as experimental subjects, it has become increasingly imperative to resort to the centuries-old practice of studying other animals in order to gain insights into our own biological makeup and to learn how to alleviate the ills that plague us. But the results of any experiment on an animal for medical purposes can be intelligible, and therefore significant, only if we know the exact normal biological properties of that animal. This means that before we can begin to experiment we must first accumulate a wealth of normal data. This time-consuming process is usually difficult to justify to the average layman who seldom concedes that it is worthwhile and who regards basic research which does not yield immediate and tangible results as unjustifiable.

To the scientist, however, biomedical experimentation can be intelligible only when the investigator is completely aware of the properties of the organisms he uses and when he uses specific organisms for specific problems. This point is worthy of belaboring and I now cite a few examples. The great discoveries in both classical and physiological genetics would have eluded us if they had not been based on previous, albeit isolated, knowledge of the peculiar giant chromosomes in the salivary glands of *Drosophila*. If we had not known the biological properties of the slime mold *Neurospora*, we might still be searching for miraculous advances in molecular biology and genetics. Pasteur could unravel the mysteries of the enzymatic processes of fermentation because he had a profound understanding of microorganisms and their chemistry. Sabin, Salk, and others contributed to the prevention and

cure of poliomyelitis only because they had at their disposal a vast store of information on viruses and immunology. Who knows how long the discovery of naturally occurring antibiotics might have been delayed if Fleming and Florey had not been astute enough to observe that the mold that contaminated their cultures had a bactericidal effect? All attempts to grow the leprosy vector, *Mycobacterium leprae*, in nonhuman primates, including chimpanzees, had failed. Now that Storrs and Kirchheimer (Storrs, 1971; Kirchheimer and Storrs, 1971) have shown that the organism that causes this ancient scourge of man can be grown in armadillos, perhaps a cure or prevention will be discovered.

The application of this principle — that specific experimental laboratory animals of known biological properties must be used to combat specific biomedical problems — by conscientious adventurers has led to the discovery of the usefulness of the owl or night monkey (*Aotus trivirgatus*) in the study of malaria, of the chimpanzee (*Pan satyrus*), squirrel (*Siamiri sciurea*), capuchin (*Cebus capucinus*), cynomolgus (*Macaca fascicularis*) monkeys and other primates in the study of atherosclerosis, of the squirrel monkey in the study of gallstones, of the stump-tailed macaque (*Macaca arctoides*) in the study of normal, spontaneous baldness, of the Celebes black apes (*Macaca nigra*) in the study of diabetes, and of aged prosimians in the study of spontaneous malignant tumors. These are not haphazard findings but the reward of systematic search, often lasting for years.

The opposite of this principle can lead to the danger of selecting a single species to be used in the study of *all* medical problems. A case in point is the use of rats, particularly in studies on sex endocrinology and behavior, and the attempt to extrapolate the findings to man. Not long ago a colleague of mine was bemoaning the results of such research and la-

menting the consequences of ignoring the fact that rat reproduction is unique to rats and that research on it is likely to
be helpful only in extending our knowledge of *rat* biology.
Beach had deplored the fact that in 1950 about 50 percent of
the published results in comparative studies of mammalian
behavior dealt with rats. Because too many endocrinologists,
biochemists, and other specialists know or care little about
natural history and comparative zoology, they are unaware of
the danger and as a result have made very slow progress in
finding solutions to our biomedical needs.

Since this principle is so clear and the results of disregarding it have resulted in wasted time and money, one wonders
why even at this late date students of animal behavior, phar-
macology, reproductive physiology, and other aspects of biomedical research still use whatever laboratory animals are
most readily available and easiest to maintain. Perhaps such a
practice stems from ignorance but more often, I think, from
a sense of urgency and expediency. Whatever the reason, the
results have been unsatisfactory. Not only was much of our
"basic" information about man in the past derived principally from observations of rats, mice, guinea pigs, rabbits, dogs,
and cats but, still more unsettling, a substantial part of our
current knowledge of organismic physiology was originally
gleaned from such dubious sources as sick and dying animals
of unknown origin and quality. Although some scientists still
regard stray dogs and cats adequate models for their pharma-
cological, physiological, surgical, and even immunological
studies, no knowledgeable contemporary investigator would
dream of using wharf rats for his controlled experiments. It is
safe to say that today every *reputable* scientist uses specific
genetic lines of rats, mice, or guinea pigs with select and predictable biological properties. Realizing that research based
on the slovenly practice of using nonselect animals is often

worthless, these scientists are insisting upon being supplied with animals of known genetic background, age, size, and medical history. Obviously, before the investigator can select the kind of subject he needs for a specific research, he must first study the biological patrimony of each species without regard to its immediate usefulness. Needless to say, it would be folly to study atherosclerosis in an animal that has never shown any spontaneous signs of the disease or the effect of chemical carcinogens in an animal immune to the agents used.

Still, I think, we need to reiterate what should have become axiomatic, namely, that the more we know about the properties, not just of a species in general, but of any one *individual* animal, the more valuable *that* animal is in research. When one reflects that no amount of careful inbreeding within a strain can produce two identical animals, one begins to appreciate the absolute meaning of "suitability" in research. The differences even between identical siblings or two animals bred from the same strain are legion. Does, for example, a litter of inbred mice all of which gestated in the same uterus, really develop in an identical environment? Certainly not. To begin with, the embryos in the uterine horns nearest the ovaries are larger and apparently better developed than their littermates and will very likely have specific survival advantages during postnatal life. Moreover, the identity of the individual placentas and the different characteristics of each part of the uterus remain unknown. The proximity of male embryos to female affects the latter, and the gender of each fetus equips it differently in its struggle to survive. In addition, even though the postnatal litter shares a mother and each neonate has access to a mammary gland, there are nutritional inequities because not all glands are the same size and not all function equally well. The unknown hazards faced by each mammal in utero and postnatally and the physical, physiolog-

ical, and psychological ravages left by these hazards are different for each individual. Experiments on drug tolerance and exposures to noxious agents call particular attention to these differences. The statistical concept of Ld 50, that is, a dose of some injurious substance lethal to 50 percent of the animals receiving it, is based upon ignorance of the principle of individual differences. When we ignore this principle, we subject the population to an experimental treatment. If all the treated subjects were really identical, all should be alive (Ld 0) or all dead (Ld 100). Even members of the same species of microorganisms exposed to the same antibiotic agents don't all die because each is a unique individual. Think of all the specific microorganisms resistant to specific antibiotics.

One more point about individual differences is that there are genetic differences in the human race between people of the same ethnic groups which are far greater than once thought. Two people chosen at random from any specific population may differ at hundreds, and possibly at thousands, of loci in chromosomes (Clarke, 1975). These important differences have affected and still affect human survival and reproduction. Some of these, no doubt, predispose individuals to different susceptibilities to disease and if one could only analyze those associations, medicine would take tremendous strides forward.

All of these digressions from the main line of argument— the search for a suitable animal model for biomedical research —have been undertaken deliberately in order to emphasize the urgency of the problem. A major step was taken by Congress over fifteen years ago with the decision to found the first primate center in 1960. Each of the centers established since that time received the same original mandate: find the most effective primates in which to study human disease and breed them as surrogates in health-related research. Responding to

that mandate, the seven regional centers have matured into highly productive scientific laboratories and have vindicated the wisdom of the decision to establish them.

Since man is a primate, it was logical to look to the many diverse species of nonhuman primates to serve as suitable substitutes for man in biomedical research. But almost simultaneously with the rising impetus to use these nonhuman primates in research came the increasing threat to their existence. Numerous factors have been in the sometimes deliberate, more often unwitting decimation of the members of this order. The worldwide and thoughtless marketing of primates for drug testing, for zoos, carnivals, and pets is probably the principal cause of their shrinking numbers. Despite criticisms to the contrary, I am convinced from long experience that far fewer wild animals have been sacrificed in the interests of scientific discovery than have been ruthlessly terminated by cruel methods of capture and the traumatic incarceration which follows; by starvation, deprivation, and crowding during transport; and, finally, by exposure to temperature extremes and to pathogenic organisms to which they have no resistance. Even when crated (often in a most barbarous fashion), many animals die en route; others die on arrival or shortly thereafter. Thus, only a few of the hundreds of animals captured for biomedical research arrive at the laboratory alive and in good condition; the rest die senselessly.

Not medical research per se, then, but man's cruel and thoughtless indifference not only to the welfare of the animals but to their economic and biological value as well has placed most primates on the list of endangered species and brought them almost to the twilight of their existence. Even the introduction of agriculture into some areas of the world where nonhuman primates once abounded has resulted in the decimation of terrestrial forms; their natural environment

was destroyed, and many had to be killed because, like their human counterparts, they were notorious marauders. Deforestation, too, has deprived many animals of the natural environments needed for survival. Many forest-dwelling primates are highly susceptible to slash-and-burn methods of lumbering, herbicides, and defoliating agents. Southwick et al. (1970) called attention to the damaging effects on Asian and South American primates of reforestation with a single species of tree such as the fast-growing eucalyptus. Many of these animals are frugivorous and granivorous and where arable lands are found live in direct competition with man. Furthermore, hungry men in many areas of the world kill primates for food. In the island of Malagasy (formerly Madagascar), for example, this practice threatens the existence of true lemurs, which are found only there.

Depletion and restriction of the environment can be calamitous to some species of animals for reasons that are sometimes obvious, but more often inexplicable (Southwick et al., 1970). For example, it is not easy to explain the virtual disappearance of a once common gibbon (*Hylobatis lar*) from China, especially since this species played a vital and long historical role in the culture of that country. All of the great apes, particularly the orangs and gorillas, face extinction. Even chimpanzees, still relatively numerous in some parts of Africa, are on the list of endangered species.

Whether a particular species of nonhuman primate is suitable for use in laboratory research depends not only on its particular biological properties, but also on abundant supply and the animal's adaptability to laboratory conditions. Recent censuses of some of the more common species of laboratory primates show an alarming drop in numbers (Southwick et al., 1970) particularly of chimpanzees, baboons, rhesus monkeys, and several other macaques. Formerly these rugged

animals were extremely numerous. As for adaptability to laboratory conditions, even at this late date, we know so little about the biological properties of many of the living primates that we cannot guess their potential value in biomedical research. Some otherwise useful animals must be ruled out for various reasons, not the least of which is the difficulty of keeping them healthy under laboratory conditions. Chimpanzees pose a hazard as laboratory animals, and the difficulties of husbandry, the near-impossibility of handling them, the dwindling natural supplies, their long gestation period, and their ontogeny severely limit their usefulness. These formidable beasts, whose anatomy and physiology most nearly resemble those of man, develop atheromata which are similar to those in human patients. Yet where is the scientist so irresponsible as to perform cardiovascular experimentation on such a rare and nearly irreplaceable animal? Apparently, the only intelligent course to follow in the biomedical studies of the great apes is to limit such investigation to observations of their blood, behavior, reproduction, and learning patterns. But despite their adaptability, chimpanzees generally become so neurotic in captivity that I find the validity of behavioral studies in this species highly questionable. Even a casual zoo goer recognizes that most caged adult chimpanzees are hysterical, have a mercurial and explosive temperament, and develop trichotylomania, that is, a compulsion to pull out and eat their own or their friends' hairs so that they become partially, if not entirely, naked.

Of all the primates in laboratory use today, rhesus monkeys are the most valuable. Until fifteen or twenty years ago, these natural citizens of India were very numerous. But according to Southwick, who with his colleagues began a ten-year study of rhesus populations in northern India in 1959,

the numerical decline in the late 1950s and early 1960s was probably due to three main causes:

1. Trapping and/or killing. Farmers, increasingly exasperated by the destruction of their crops by monkeys, resorted to trapping or killing them. In the late 1950s, more than 100,000 were exported every year and later more than 200,000. Juveniles suffered the greatest depletion (Southwick et al., 1970).

2. Environmental changes. The census takers observed that land changes, principally deforestation, created an unfavorable environment for rhesus monkeys. The steady decline that began in the early 1960s stabilized around 1966, partly because of the accelerated move of human populations to the town and city, partly because of curtailed exports to fewer than 50,000 per year.

3. Disease. Recent data show a decline in rhesus population in villages and rural areas, stable levels in forest populations, and increases in certain urban areas. Unfortunately, however, the urban monkeys are mostly in poor health, suffering high incidences of respiratory and enteric diseases. Obviously, such animals are poor risks for laboratory experimentation.

Planned management of rhesus populations in India would supply enough animals for biomedical purposes worldwide, provided, of course, that the harvesting of these animals was properly controlled. But even were such a program to be initiated, the monkeys' adjustment to changes in habitat through exportation is often unpredictable. Moreover, India's traditional practice of protecting monkeys has often deterred exporters, and the exportation of primates has always been resisted by the governments of India, South America, and Malagasy. In fact, official restrictions have become so tight and

animals so expensive that legitimate dealers are hard put to stay in business. The result has been a flourishing black-market trade, beyond the control of both governmental and scientific jurisdictions. As a result of this illicit traffic in lemurs, Celebes black apes, orangutans, and gorillas, the number of feral animals available to scientists continues to decline.

The time is long overdue, therefore, to heed the mandate to initiate sound breeding programs in the United States, not only by primate centers but also by private individuals for commercial sale under the guidance of experts. But such a vast program of breeding cannot be undertaken on a national scale until we have acquired more complete data on natural primate populations: rhesus monkeys, crab-eating macaques, squirrel monkeys, owl monkeys, marmosets, pig-tailed macaques, gibbons, African vervets (including the talapoin monkeys), the great apes, and the other primate species already in use. Enough knowledge has already accrued about the reproductive biology of some of the more useful laboratory primates to produce breeding colonies that are enabling endangered species to gain a new foothold on survival. But we have to have more ecological studies to provide the information that is needed for the successful breeding of animals in captivity. Too many of the field studies of primates during the last fifteen or more years have concentrated on behavioral observations of feral animals to the detriment of studies on primate population ecology (Southwick et al., 1970). Fortunately, the breeding experiences of the seven primate centers in the United States are available to all who have the capital to raise primates for financial profit. We are hopeful that ultimately these two sources of increasing primate populations will provide solutions to a problem that at the moment appears almost unsolvable.

Having established what appears to me to be an incontro-

vertible fact — that where ethical and moral considerations preclude the use of human beings as experimental subjects, nonhuman primates are in most cases the most feasible substitutes — I am at a loss to understand the continued opposition to their use among the general public. In a free society which abhors the use of any of its members as experimental subjects, nonhuman primates are indispensable for the solution of many of that society's medical problems. For example, in studies of fetal and neonatal respiratory distress syndrome, the dynamics of amniotic fluid, and the possibly beneficial effects of glucocorticoids, monkey fetuses are excellent substitutes for human ones because their placentas are as permeable to glucocorticoids as those of man. Moreover, because the hemochorial intervillous circulation of human subjects and that of monkeys are structurally and functionally similar, monkeys are invaluable in studies of the regulation of uteroplacental blood flow. They are also excellent models for studies of premature labor and the regulation of myometrial activity.

Why then the opposition? Predictably and frequently, scientists are challenged and will continue to be challenged to justify their use of animals in experimentation. Many animal lovers demand, often justifiably, to know whether animals are being treated humanely. I agree that scientists should be prepared to account for their treatment of their animals since we know that one can very easily become callous to suffering, whether in man or in nonhuman animals. Still, I wonder whether many of the general public are truly qualified to pass judgment on these matters. Too many of them are still entrenched in the Romantic trend of idealizing animals which began in literature over a hundred years ago and has found its most saccharine expression in the personifications of the Walt Disney productions and their many imitators. From child-

hood on, we have been conditioned by the media—literature, the comic strips, television, the cinema—to idealize animals, to accept fallacy as gospel, and as a consequence to agonize unduly over the welfare of animals.

Knowing what he knows and secure in his knowledge of his own laboratory conditions, the scientist can be forgiven if he sometimes becomes impatient with the oft-repeated harangue that "suffering" caged animals should be given their freedom. Ironically, such concern, often more anthropomor-

Fig. 2. Inbred malformations in dogs. In the upper row are (left to right) the head of a bloodhound and two views of a pit bulldog; in the center, a greyhound; and at the bottom are a dachshund (left) and a naked Chihuahua (right).

Fig. 3. A prize-winning bulldog showing the whole gamut of inbred malformations.

phic than rational, is not accorded all animals. Sometimes the very people who anguish over the suffering of an experimental cat, dog, or monkey indulge themselves in the "sport" of hunting and fishing and willfully participate in the destruction of such subjectively unattractive animals as toads, snakes, rats, mice, and such alleged vermin as wolves and coyotes. People who purport to be fond of domestic pets would do well to reflect on the barbarisms they themselves perpetrate or allow to be perpetrated on these very pets by selecting genetic defects and breeding for them. Think, for example, of the achondroplastic condition of bulldogs, basset hounds, bloodhounds, and dachshunds (Figure 2); the deliberate diminution of the skulls of collies, Shelties, and Irish wolfhounds; the nakedness of Chihuahuas, the deafness of Dalmations and white Angora cats, and other equally barbaric, deliberately induced deformities! A syndrome of defects is evident in the short-faced breeds, bulldogs (Figure 3), mastiffs, and boxers: dental malocclusion, olfactory difficulties and disrupted lacri-

mal glands, and oversized tongues. In these faces I see concentrated the dog fancier's senseless cruelty. It should be of some interest that nearly all household dogs which are not allowed to roam free suffer from prostatic hyperplasia. It should help to balance the criticism leveled against scientists to reflect that all feral and most stray pets live in constant danger from natural enemies and that their actual life expectancy in the wild is only a small fraction of their potential expectancy. Wild macaques at fifteen are often prematurely aged, emaciated, almost toothless, and sometime brutally scarred whereas those born in captivity or caged when young are still healthy and vigorous at the same age. Many captive songbirds live from ten to fifteen years whereas free yearlings of the same species have a mortality rate of more than 60 percent. And who can deny that caged birds, though fidgety, enjoy better nutrition, care, and health than wild ones? Unhappily, they are too often allowed to die of old age; the sight of birds in poor feather, slow moving, arthritic, and no longer able to sing is not a pretty one.

Animals are used in biomedical research simply and solely to extend our knowledge and deepen our understanding of nature, especially of human nature, in health and disease. But biomedical experimentation is only one facet of a much wider exploration; its value cannot, then, be assayed in isolation. Thus, if the use of animals in medical research is made extremely difficult, curtailed, or actually prohibited (as we are often warned and threatened that it will be!), the loss to mankind will not be confined to the physiological aspects of human life. Still, despite current difficulties and opposition, I predict that the next few years will witness the increasing use of experimental animals, particularly of primates. I will even venture to assert that once animal experimentation has peaked, it will decline because, as Medawar (1973) has ob-

served, research on well-selected, adequate experimental animals eventually yields the kind of knowledge which may some day make it possible to dispense with them altogether.

As long as man lives, he will have health problems, and there will always be a need for medical research. Problems related to reproduction and its control, problems associated with the degenerative changes of old age will persist whether we cure cancer and heart disease or not. In fact, the added longevity which is a dubious bonus of such cures merely adds to these other problems. Some nonhuman primates, because of their biological properties, are especially useful in the solution of these problems. I shall attempt in later chapters to take a closer look at the members of this, to us, extremely valuable mammalian order. Because of the extraordinary differences among species, I have chosen to make these discussions comparative. But to have meaning and focus, comparative studies must have a specific point of reference. Since man is the most advanced of the primates, I shall deal with the properties of the other members as they pertain to him.

CHAPTER 2

THE NATURAL HISTORY OF PRIMATES

The name *Primate* embraces a wide assortment of different animals which have in common a number of generalized and specialized characteristics —structural, physiological, and behavioral. Consider, for example, the diminutive dwarf galagos, pygmy marmoset, and mouse lemur (Figure 4) whose size belies their common ancestry with man and the gorilla, which sometimes exceed 300 and 600 pounds respectively. Most scholars agree that because of their generalized biological attributes these greatly diversified animals share the classification Primates. In their probable evolution from a single precursor stock, they have retained ancestral characteristics that have been modified in some and specialized in others. Some, perhaps including man, tend consistently toward specialization, while at the same time preserving various primitive ancestral features. If we avoid the minutiae of generalization and specialization, which I insist are relative, we can attempt to analyze these primate attributes. We can also per-

NOTE: This chapter is based on a lecture in the Wesley W. Spink Lectures on Comparative Medicine presented at the University of Minnesota, St. Paul campus, on October 14, 1975.

Fig. 4. A mouse lemur, the smallest prosimian. (Courtesy of San Diego Zoo.)

haps find answers to our questions about primate behavior. Why, for example, do man, gorilla, chimpanzee, baboon, macaque, and patas monkey prefer a terrestrial life when most members of the order are arboreal? Since most primates survive on a vegetable or omnivorous diet, why are tarsiers completely carnivorous, and lorises, some galagos, and a few simians partially carnivorous? Why are pottos and lorises slow-moving when most primates are alert and quick? Why, finally, are some social, others solitary; some vociferous, others taciturn?

Before we can begin to answer these questions, we must first establish what primates are. To do so is no simple task. Systematics is a delicate and complex art which should not be undertaken lightly by the unwary amateur. To define something, one must first name it. And there's the rub! For as Simpson warned over thirty years ago, to define the Order Primates is to court confusion, and the closer we come to defining man, the more confused we become. As a consequence, the taxonomist traditionally is at the vortex of the storm, for it is he who gives names and assigns individuals to categories. So violent is the controversy over classifying primates that even today some genera and species are given different names by different taxonomists. For example, the Chacma baboon of South Africa is variously called *Papio comatus*, *P. ursinus*, and *P. porcarius*, and the Asian macaques have been identified under at least twenty scientific names! The East African baboon, the mandrill, the Celebes black ape (which, incidentally, is not an ape but a macaque), and the chimpanzee are even assigned different genera by various systematists. Such taxonomic disagreements reflect not so much muddleheadedness and scholastic hairsplitting as the near impossibility of arriving at a consensus on the classification of a varied and complex group of animals about which biological data—immunological, behavioral, and evolutionary—are still being amassed. To be meaningful, of course, such data must be linked to genera and species about whose names there can be no controversy. That we are still far from such an agreement may come as a surprise or even shock to the layman and fledgling taxonomist!

The shock is understandable when we consider that the ancestry of contemporary primates was already distinct in the Eocene period, some 50 million years age (Table 1; Figure 5). The catch is that during the ages-long adaptation to arboreal

Table 1. Geological Periods of Quaternary and Tertiary Eras

Geological Periods	Started
Holocene (wholly recent)	10,000 years ago
Pleistocene (most recent)	2-3 million years ago
Pliocene (more recent)	13 million years ago
Miocene (less recent)	27 million years ago
Oligocene (few recent)	38 million years ago
Eocene (dawn of recent)	56 million years ago
Paleocene (old of recent)	66 million years ago

life, primates were evolving into many forms which in turn evolved into still other forms during their adaptation to a bipedal life on the ground. The temptation to link evolutionary history to structural adaptation is a fascinating one. For example, the fact that many nonhuman primates can widely abduct their powerful great toe suggests that during the adaptation these arboreal primates lost the prehensility of their great toes which had to perform new functions and to adapt to terrestrial support and bipedal walking.

As we have seen, the evolutionary history of primates is a long one. Primates became separated from other mammals near the beginning of the Age of Mammals and can be distinguished, among other features, by an enlarged brain and adaptation to arboreal life. Whether the lower species of primate, the Prosimii, qualify as members of the order is debated by some. Certainly, tree shrews, lorises, galagos, lemurs, and tarsiers are a motley assortment and only vaguely resemble what we generally recognize as primates. The tree shrews (Family Tupaiidae) from Southeast Asia, for example, are so "primitive"* that some zoologists still regard them as members of the Order Insectivora (Figure 6). To include them in the Order Primates, we have to overlook the absence of such

---

*The word *primitive*, like *unspecialized*, is so difficult to define that almost everyone who uses it is either purposefully or unconsciously vague about its meaning.

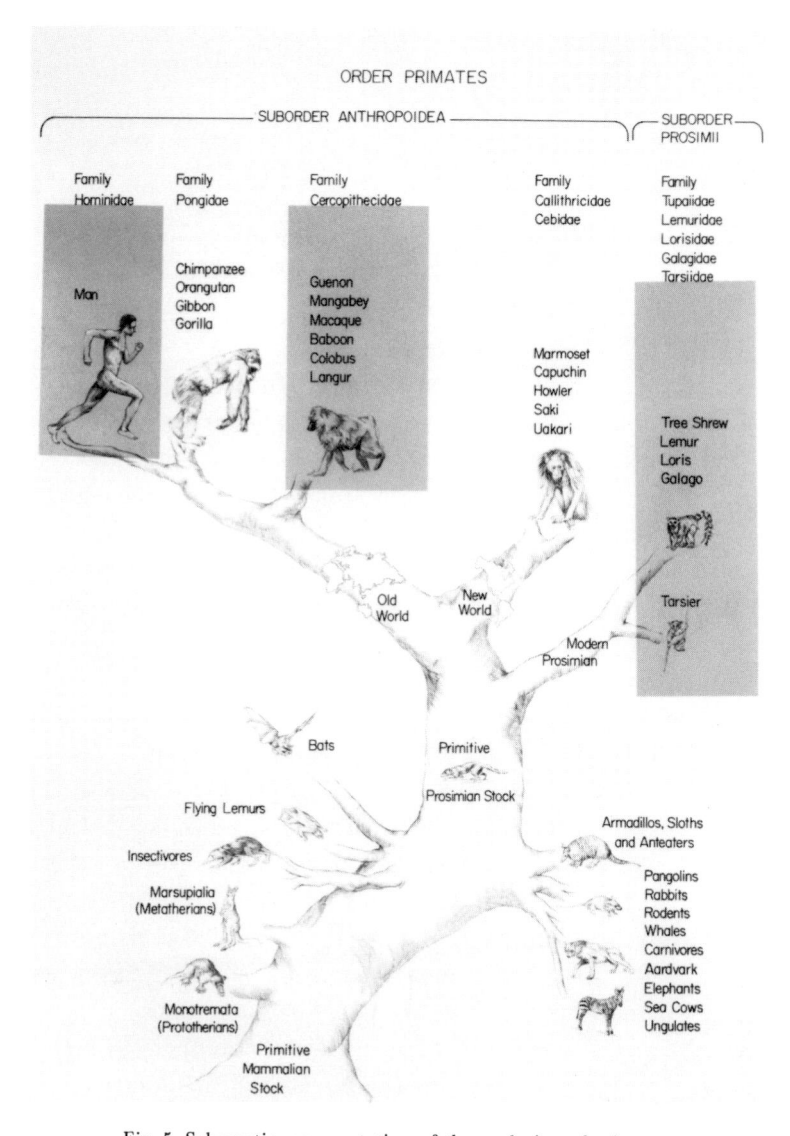

Fig. 5. Schematic representation of the evolution of primates.

Fig. 6. Tree shrew (*Tupaia glis*), a primate of dubious standing. a. The entire animal. (Courtesy of San Diego Zoo.) b. Detail of the head.

commonly accepted simian characteristics as nails instead of claws, few teeth, a shortened snout, enlarged brain, highly developed vision and diminished sense of smell, reduced litter size, and a unique manner of rearing the young. Although the skulls of the tree shrews are not like those of primates, their behavioral locomotor, and immunochemical attributes are close enough to warrant their inclusion. Moreover, all the evidence—paleontological, anatomical, locomotor, behavioral, and immunological—suggests that they are either primates or the closest nonprimate allies alive. At any rate, they do provide at least a clue to the kind of animal primates evolved from because apparently their ancestors developed into lemurlike animals some 50 million years ago. Today the three families of *true* lemurs are found only in Malagasy. Although their primate characteristics may not be immediately apparent (Figure 7), their anatomical features are distinctly simian, even though some of them are also found in nonprimate mammals. The glabrous (hairless) moist nose and projecting snout, the midline cleft in the upper lip, the usually large mobile ears, and the three pairs of nipples are more characteristic of carnivores than of simian primates. Hence, earlier German zoologists referred to lemurs and other prosimians as *Halbaffen* or half-monkeys. Some anthropologists have theorized that soon after their accidental colonization of Malagasy, lemurs became segregated from the rest of the world's primates. Writing about these small animals, anthropologist J. Buettner-Janusch stated: "They are a unique experiment in primate evolution. If we are clever enough to take advantage of this experiment, we may continue to learn more about the nature of the mammalian order to which we belong and more about evolutionary processes in general."

The most widely distributed prosimians are the tree-dwelling, nocturnal, and solitary Lorisidae: galagos, pottos, ang-

a

b

c

d

Fig. 7. a. Ringtailed lemur (*Lemur catta*). b. White-fronted lemur (*Lemur fulvus albifrons*). c. Red ruffed lemur (*Lemur variegatus*). d. Sifaka (*Propithecus*). (Courtesy of San Diego Zoo.)

wantibos, and lorises (Figure 8). Galagos (also known as bush-babies), pottos, and angwantibos are found in nearly all parts of Africa south of the Sahara; lorises in parts of India, Burma, and the Malayan Peninsula, Ceylon, and the islands off south-eastern Asia. Together these four groups form two distinct subgroups: the pottos, angwantibos, and lorises are slow creepers and climbers with short tails and small ears; the gala-gos are quick saltatory animals with long bushy tails and large ears.

Today, the once-large group of tarsiers is represented by the single genus *Tarsius* (Figure 9). Fifty million years ago many different forms and sizes of this genus were found throughout Eurasia, Africa, and North America, but today they are confined to the islands of the East Indian Archipela-go from the southern Philippine and Celebes islands to Su-matra. Roughly the size of a rat, *Tarsius* has several unique anatomical features, including large eyes, well-developed nic-titating membranes, long hind limbs that enable it to take prodigious leaps in any direction; large, constantly twitching ears with an undulating edge, and a remarkable way of rotat-ing its head which enables it to look around straight behind. In general, however, the anatomy of the nose, certain repro-ductive organs, and the brain is distinctly simian. Whether the tarsier ancestors were in direct lineage to the prosimians from which the Anthropoidea diverged is not certain, but in spite of those who question it, such appears to be the case.

The simian primates—monkeys,* apes,† and man—belong to the suborder Anthropoidea. The first fossil anthropoid re-mains date back some 38 million years to the Oligocene pe-

---

*The word *monkey* is a distortion of *homunculus*, the generic name for an ex-tinct form of New World monkey from the Miocene period which flourished some 28 million years ago.

†The English language, to my knowledge, is the only one that has different names for monkeys and apes; other languages lump them under one name: *scim-mia* (Italian), *Affen* (German), *singe* (French), etc.

Fig. 8. a. Potto (*Perodicticus potto*). (Courtesy of San Diego Zoo.) b. Fat-tailed galago (*Galago crassicaudatus*). c. Angwantibo (*Arctocebus calabarensis*). d. Loris (*Nycticebus coucang*). (Courtesy of San Diego Zoo.)

Fig. 9. Two views of a Philippine tarsier (*Tarsius syrichta*). (Courtesy of San Diego Zoo.)

Fig. 10. Pygmy marmoset (*Cebuella pygmaea*). (Courtesy
of San Diego Zoo).

Fig. 11. Two views of golden lion marmosets (*Leontideus rosalia*).
(Courtesy of San Diego Zoo.)

riod. Of the many genera that evolved, only about forty are extant, most of them on the endangered species list. As their movements, facial expressions, and ready response to their environment attest, most of the anthropoids are not only alert and active but adventurous and inquisitive as well. The torpid ways of orangs contrast to this curiosity and seem to belie their intelligence.

The suborder Anthropoidea is divided into six families: the New World (platyrrhine) Callithrichidae (marmosets) and Cebidae (squirrel monkeys, spider monkeys, etc.) of Central and South America; the Old World (catarrhine) Cercopithecidae (macaques, baboons, etc.), Hylobatidae (gibbons), Pongidae

Fig. 12. Two views of a night or owl monkey (*Aotus trivirgatus*). (Courtesy of San Diego Zoo.)

Fig. 13. The red uakari (*Cacajao*) from South America. Note
the short hair on the entire head. (Courtesy of
San Diego Zoo.)

(great apes) of Africa and Asia, and Hominidae (man). The
smallest of the Callithrichidae, the pygmy marmoset (*Cebuella pygmaea*), is about six inches long (Figure 10); the largest,
the golden lion marmoset (*Leontideus rosalia*), about seven-
teen inches long (Figure 11). These animals have claws instead of nails on all digits except the big toe. The family Cebidae comprises the Atelinae (spider and woolly monkeys), the
Alouattinae (howler monkeys), the Aotinae (night monkeys

and the dusky titi) (Figure 12), and the Pitheciinae (the bizarre sakis and uakaris) (Figure 13). They are believed to have evolved from an offshoot of the main evolutionary line of descent and to have developed independently and in the relative isolation from the South American land mass. Howler, spider, and woolly monkeys have a true prehensile tail with the ventral skin modified as a friction surface, identical with that of the hands and feet (Figure 14). Capuchins, of the family Cebidae, which also curl their tails around branches, lack such anatomical differentiation, and squirrel monkeys scarcely use their tail at all for prehension. Platyrrhines possess other characteristic features, such as the hollowed-out hyoid bone in the upper part of the howler neck beneath the

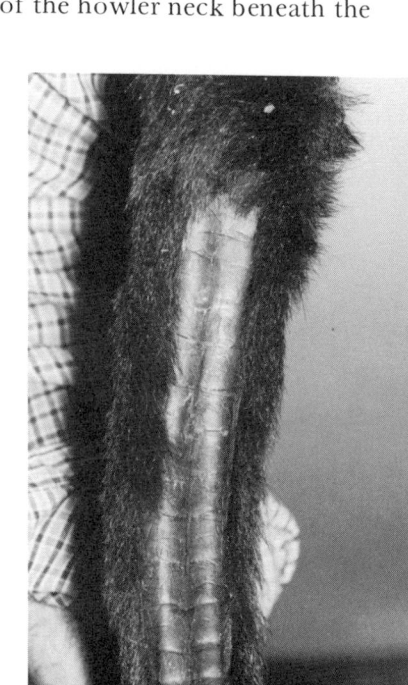

Fig. 14. The woolly monkey (left) and its prehensile tail (right).

Fig. 15. A proboscis monkey (*Nasalis larvatus*) on the left and a Douc langur
(*Pygathrix nemaeus*) on the right. (Courtesy of San Diego Zoo.)

Fig. 16. A DeBrazza's guenon (*Cercopithecus neglectus*) from
Africa. (Courtesy of San Diego Zoo.)

Fig. 17. A number of patas monkeys (*Erythrocebus*). (Courtesy of San Diego Zoo.)

Fig. 18. Hamadryas baboons (*Papio ?*). (Courtesy of San Diego Zoo.)

Fig. 19. Gelada baboon (*Theropithecus*). (Courtesy of San Diego Zoo.)

tongue which enables the animal to produce his explosive booming sounds.

The more than seventy species of the large family Cercopithecidae are divided into two subfamilies: (1) the Colobinae, e.g., the colobus monkey of Africa and the langurs and the proboscis monkey of Asia (Figure 15), and (2) the Cercopithecinae, e.g., the mangabeys (*Cercocebus*), guenons (*Cercopithecus*) (Figure 16), patas (*Erythrocebus*) (Figure 17), macaques and Celebes black apes (*Macaca*), mandrills (*Mandrillus*), baboons (*Papio*) (Figure 18), and gelada baboons (*Theropithecus*) (Figure 19). (Some systematists, "lumpers,"

Fig. 20. A siamang (*Symphalangus*) on the right and a crested gibbon (*Hylobates hoolock*) on the left showing the inflated gular pouch. (Courtesy of San Diego Zoo.)

place mandrills and all the baboons in the single genus *Papio*.) The Colobinae are preferentially leaf-eaters and arboreal although one genus (*Presbytis*) is partially terrestrial. Even though these Old World monkeys are less well adapted for arboreal life than New World monkeys, macaques, baboons, and patas monkeys are skillful climbers when they choose or have to be; generally, however, they remain terrestrial during the day.

The two families of apes, the Hylobatidae (Figure 20) or lesser apes and the Pongidae or greater apes, are grouped together by Simpson in the superfamily Hominoidea. The former consist of gibbons (*Hylobates*) and siamangs (*Symphalangus*); the latter, of orangutans (*Pongo pygmaeus*), two species of chimpanzee (*Pan troglodytes* and *P. paniscus* Figure 21, a

nearly extinct pygmy form), and gorillas (*Gorilla gorilla*). Although small (15 to 20 pounds with a brain weight around
100 grams), gibbons are the most numerous and successful of
the apes. Once abundant in China, they are now found only
in Malaya and parts of Southeast Asia. Using their long slender arms, they swing through the forest canopy like trapeze
artists. When they descend to the ground, they usually walk
bipedally and upright, holding their forelimbs out for balance.
An inflatable sac beneath the chin of the siamangs (the larger
gibbons) acts as a resonating chamber which greatly amplifies
their whooping cries and the haunting wail that sounds like a
hornpipe and seems not to come from the animal at all.

Fig. 21. Two views of a pygmy chimpanzee (*Pan paniscus*).

a

b

c

d

Fig. 22. a. Male gorilla (*Gorilla gorilla*). (Courtesy of San Diego Zoo.) b. Male (left) and female (right) showing the difference in size. (Courtesy of San Diego Zoo.) c. Portrait of a male. d. Portrait of a female.

Fig. 23. Male (right) and female orangutans (*Pongo pygmae-us*). (Courtesy of San Diego Zoo.)

Despite obvious differences in gross structural features, size, and behavior, gorillas and chimpanzees, which inhabit the tropical forests of Africa, are thought by some anthropologists to be close enough to belong to the same genus. (The few remaining pygmy chimpanzees live south of the Congo River.) Gorillas are the most massive of the extant primates, full-grown males often weighing up to 600 pounds (Figure 22). Orangs, which inhabit Borneo and Sumatra, weigh up to 170 pounds, most of which is "blubber," whereas adult chimpanzees weigh from less than 125 pounds. Because of their size and structural features, adult chimpanzees and gorillas are less at home in the trees than gibbons and therefore prefer a terrestrial existence; orangs (Figure 23), on the other hand, are mostly tree dwellers and excellent brachiators.

All of these apes have developed various walking techniques. Because chimpanzees and gorillas rest the forepart of their body on the dorsal second phalanges of digits, 2, 3, and 4 instead of on their flat palms, they are called knuckle walkers (Figure 22). The dorsal skin of the phalanges that contact the ground is a specialized friction surface. Orangs walk on the inside of clenched fists and lack this special feature.

Unlike Old World monkeys, the great apes have no calloused sitting pads (ischial callosities) and no external tail. Thus, their pelvic basin and perineal area is structurally like man's. Their reproductive processes, too, resemble man's. Unlike human infants, their newborn are well developed, but like them they are carefully reared for many months postnatally. They reach puberty later than monkeys, between six and eight years of age.

The family of man includes both the extinct and the extant races of Hominidae. All the evidence — anatomical, behavioral, biochemical, and serological — relates man to the apes. For example, tests of the reactions of blood proteins of gorillas and chimpanzees show them to be closer to those of man than to those of gibbons and orangutans! The results of these studies, however, have led to conclusions about the timing of the evolution of man and apes that are consistent with those deduced from the fossil record. Since the soft parts of ancient men and apes could not be preserved, only the fragments of skeletal structures are available for study. Although such paleontological evidence is distressingly incomplete, new fragments are being discovered almost daily. Let us digress briefly to search the record of human evolution left by these fossil fragments. Without such reference to these fossil records, I doubt whether man's relationship to living primates or his phylogenetic history can be fully appreciated.

The earliest primate fossils, found in the middle Paleocene

deposits of North America, are those of prosimians. During the late Paleocene and the Eocene period (40 to 60 million years ago), these prosimians expanded their range and became extensively diversified in both North America and Europe. Many early members acquired enlarged, rodentlike anterior teeth; and the extinct families — Carpolestidae, Plesiadapidae, and Phenacolemuridae — may well have occupied the niche later taken over by the true rodents at the end of the Eocene. In fact, the only link between these mammals and the later primates is the tenuous evidence of similarly shaped molars. The number of rodentlike families of prosimians probably peaked at the beginning of the Eocene epoch and in North America and Europe diversified into fifty or sixty genera; from then on, they were gradually outnumbered by more modern forms of primates. The striking similarity between many of the American and European forms during the late Paleocene to the middle Eocene suggests an extensive faunal interchange. By the late Eocene, however, the similarity between the populations of the two areas gradually weakened, probably because faunal interchange had been interrupted, and thenceforth New and Old World primates appear to have evolved independently.

Either because of climatic changes or because of an evolutionary expansion of the rodents, the prosimians were forced to migrate south and their populations were reduced. Whatever the cause, by the end of the Eocene, prosimian fossils were rare. This scarcity constitutes one of the most serious evolutionary gaps of the primate order because it occurred when Old and New World monkeys were evolving from their prosimian ancestors. As a result, none of the later fossils or extant species can be directly traced to the numerous genera and species from the previous periods.

Except for the well-preserved skull of *Rooneyia* from Tex-

as, prosimian fossils had virtually disappeared from the Oligocene record (ca. 28 to 40 million years ago) in America. The
characteristic features of New World monkeys had not evolved
in this animal, which is clearly a North American Oligocene
prosimian. Lacking fossil evidence, we can only surmise that
the crucial transition from prosimian to monkey took place
somewhere in the southern United States, Mexico, or Central
America. Yet another attractive possibility is that some seed-
ing or colonization took place in South America from Africa,
when the two continents were much closer. Similarly, Malagasy lemuriforms seem to be tied to the African Lorisidae
through the minuscular *Microcebus* of Madagascar and the
similarly small *Galago demidovii* of West Africa. Although
both *Cebupithecia* and *Homunculus*, which had evolved
in South America by Miocene times (12 to 28 million years
ago), resemble modern monkeys, they are completely differentiated and provide no evidence of the transition from prosimian to monkey in the Western Hemisphere.

No unequivocal link can be forged between extant species
of monkeys and apes, the fully evolved and diversified catar-
rhines (which include a number of forms from the Oligocene
period), and the ancestral fossil prosimians. Apparently during the late Eocene and early Oligocene, the shrinking prosimian populations were more and more restricted geographi-
cally, perhaps because of competition with the expanding
populations of Old World monkeys. The patterns of contemporary distribution show that only such nocturnal prosimians
as bushbabies (*Galago sp.*), angwantibos (*Arctocebus calabarensis*), pottos (*Perodicticus potto*), and lorises (*Loris tardi-
gradus, Nycticebus coucang*) survived cohabitation with the
diurnal Old World monkeys. All extant diurnal prosimians
are found on the island of Madagascar where apparently monkeys never existed.

The evolutionary divergence of monkeys and apes can be traced in the Miocene and Pliocene deposits of Europe and Africa (Table 1). The living colobine monkeys of Africa and Asia are definitely related to *Mesopithecus* of the European Pliocene (2 to 12 million years ago), and the early evolutionary stages of the gibbon can be traced to the *Pliopithecus* remains from Miocene strata in Africa and Asia. Moreover, on the basis of cranial and dental structure, as well as of foot and hand, modern hylobatids are closely related to *Pliopithecus*. However, the modified arm length and the type of joints characteristic of the former, which had only begun in *Pliopithecus*, differ from those of monkeys. Both Proconsul (so named because its teeth and jaws resembled those of Consul, a chimpanzee in the London Zoo) and *Dryopithecus* (tree ape) are thought to be probable links with the modern chimpanzees and great apes, but Proconsul possessed so few unequivocally human features that its ancestry to man is disputed, and the fossil remains of the latter are too fragmentary to establish them as ancestors of the modern great apes.

Primates are characterized not only by numerous generalized and almost primitive mammalian features but by highly specialized ones as well. They retained a generalized limb structure which enabled them to move their limbs freely, to grasp objects, and to depend on the mobility and strength of their limbs and the girdles that supported them. Most terrestrial primates retained and even refined the prehensility of their hands and feet. The freedom of the digits is most marked in the thumbs and big toes, at least one of which is opposable in most forms, that is, can be rotated or turned around on its axis so that its palmar surface can be brought against that of any of the other fingers. Dating back to very ancient times, opposition probably indicates an ancestral adaptation to an arboreal existence. Certain extant arboreal marsupials, for ex-

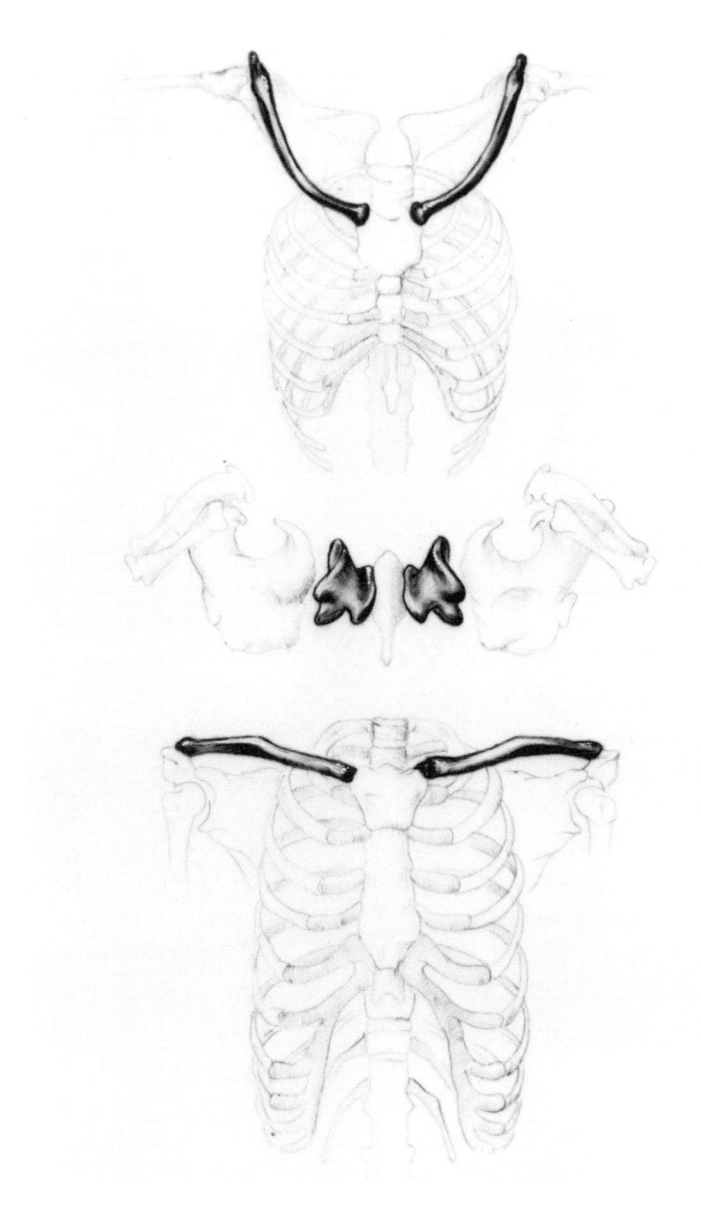

Fig. 24. The clavicles of a gibbon (top), a mole (middle), and a human (bottom).

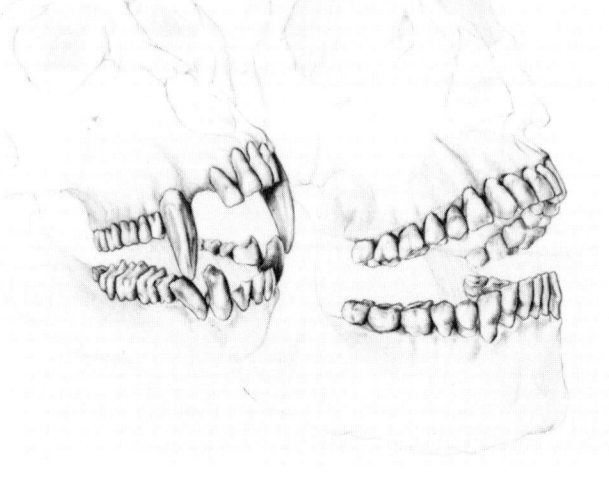

Fig. 25. A comparison of the teeth of a rhesus monkey (left) and those of a man (right). Note the difference in the size of the canines.

ample, can also oppose their digits. The freely opposable digits of primates are characterized by broad flat nails instead of the more primitive claw. Tree shrews have claws on all digits; the second toe of pottos, lorises, and galagos retains a claw whereas the unusual nocturnal lemur *Daubentonia* (aye-aye) has claws on all but the big toe, which has a nail. The nails of lemurs are keeled and pointed like claws.

All primates have retained another generalized and primitive anatomical feature which is indispensable for arboreal life, namely, the clavicle, which provides a well-strutted attachment of the upper limbs to the trunk, a distinct advantage to animals that swing by their arms. When man became bipedal and upright, the clavicle attained new significance by providing an important strut that enabled him to move his upper limbs in space (Figure 24).

Arboreal life not only involved a modification of limbs and musculature but also made available items of diet not easily obtainable on the ground. Most primates except man feed largely on young shoots, leaves, and fruits without, however, excluding insects and small animals. Since most living simian primates are omnivorous, their dentition has retained such primitive mammalian features as numbers, cusp pattern, and functional categories. Except where the fundamental pattern is modified and specialized, primate dentition consists of small incisor teeth, variously developed canines, and moderately developed premolar and molar teeth for grinding and crushing (Figure 25).

Except for the museum-styled systematists, primatologists have rarely been concerned with the widely diverse kinds of pelage, cutaneous appendages, and coloration that characterize primate skin. For that reason, the skin is a rich untapped source of biological information which is as varied as the behavior of the living animals. Because the differences are so numerous, only a few of the most common and obvious ones will be discussed here.

Among these is the wide range of color. In heavily furred animals, often only the hairs are pigmented. Celebes black apes, however, have a heavily pigmented epidermis but a nonpigmented dermis; rhesus monkeys have just the opposite. Sometimes the epidermis of chimpanzees contains much melanin, sometimes none. Melanocytes were observed in the epidermis of all nonhuman primates studied by my colleagues, but sometimes, as in rhesus monkeys, they are mostly amelanotic.

Except for certain kinds of nevi, human dermis has few or no pigment-containing cells; the pigment is usually restricted to the epidermis and hair follicles. In nonhuman primates, however, the number of such cells ranges from few to many,

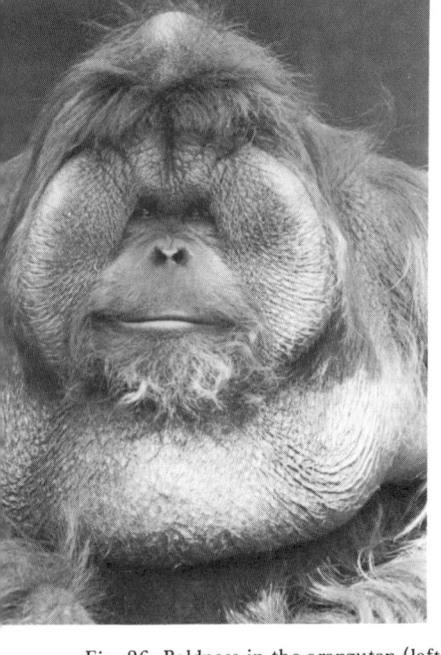 

Fig. 26. Baldness in the orangutan (left) and a male stump-tailed macaque (right). See also Figure 13.

depending on the species. They are found in the papillary or reticular layer of the dermis. In some spider monkeys, the dermis has so many pigment cells that it seems abnormal. Several species, including chimpanzees, lemurs, and macaques, have distinct perivascular and perineural melanocytes in the dermis.

Contrary to popular belief, man is not the only mammal who becomes bald. The uakari (*Cacajao*), the stump-tailed macaque (*Macaca arctoides*), the orangutan (Figure 26), and the chimpanzee have a fully or partially "naked" forehead, which is a kind of baldness (Montagna, 1972). During preadolescence, their forehead (like that of human fetuses during late gestation) and scalp are hairy. In preadolescent uakaris from South America, the long, sparse hairs on the scalp and

forehead, are replaced by short hairs, first on the forehead and then over the entire scalp. This process, which has been studied particularly in stump-tailed macaques, is exactly comparable to that in some balding men.

The numerous sebaceous glands which characterize human skin are found only in some other primates. Though smaller in all simian nonhuman primates than in man, they are largest and most numerous on the face and anogenital surfaces.

Large sebaceous glands on the buccal and oral mucosa, and occasionally even on the gums and tongue, are unique to adult human beings; diligent search of the same sites in other primates, including the great apes, has failed to show similar glands.

Most mammals have apocrine glands over their hairy skin, usually associated with hair follicles. The eccrine glands found in a few species are restricted to the glabrous areas. In the hairy skin of the more advanced primates, however, the ratio is reversed, the apocrine glands becoming sparser as the eccrine glands increase. Unnecessary to the biological economy of modern man, apocrine glands are restricted to specific areas of the human body whereas eccrine glands are found almost everywhere on human skin. Chimpanzees and gorillas have more eccrine than apocrine glands; the great apes have about equal numbers of each.

In fetal man, some other primates, and some domestic animals, apocrine glands develop in close association with hair follicles and, in adults, open inside the canal of the follicle. In many adult animals, including most nonhuman primates, the glands open directly onto the surface—near to, but not inside, the follicle orifices. Apocrine glands may even open onto glabrous surfaces, as in the ring-tailed lemur. Thus, these glands do not necessarily develop from and remain associated with hair follicles.

In a human fetus, rudiments of apocrine sweat glands appear nearly everywhere on the body at five to five and a half months. After a few weeks, however, most of them disappear except in the external ear canal, the axilla, and occasionally around the navel and anogenital surfaces. In adult human beings, single glands can be found almost anywhere. The ancestors of man may have had apocrine glands widely distributed over the body; if so, the embryonic rudiments may be reminders of a once widespread organ system. In the specific areas mentioned above, the glands are very large and numerous; for example, they are so large in the axilla that the coils press upon each other, forming adhesions and cross-shunts of such complexity that the glands are more spongy than tubular. The complex of these large apocrine glands, together with an equal number of eccrine glands, forms what is known as the axillary organ. Found only in man, the gorilla, and the chimpanzee, this organ accounts for the similar axillary odor in all three species.

Like their palms and soles, the lower surface of the prehensile tail of spider, woolly, and howler monkeys from South America is differentiated, even with dermatoglyphic imprints, and has numerous eccrine sweat glands. The gorilla and the chimpanzee, which are knuckle walkers, also have the aconal skin of the knuckles specialized like friction surfaces and abounding in eccrine glands.

Significantly, the eccrine sweat glands on the hairy skin of nonhuman primates appear to be structurally perfect; but if they sweat at all, they do so minimally, even when injected with potent sudorific drugs. The glands on the palms and soles, however, sweat profusely and, like those of man, respond to either cholinomimetic or adrenomimetic drugs. Nerves have been demonstrated which contain acetylthiocholinesterases and catecholamines. Regardless of heat levels in

the environment, the skin of monkeys remains mostly dry. The only sweating I ever saw was in a rhesus monkey dying of cardiac failure and in a gibbon under deep sedation.

Since the skin of primates possesses all of the known features found in other mammals, it can be considered generalized but also specialized in certain regions of some primates.

One of the most important and characteristic of the primate specializations, as indicated earlier, is the relatively large size of the brain in proportion to the size of the body, which is mainly due to the increased mass of the cerebral cortex, and a corresponding progressive enlargement of the optic lobes of the brain. As a result, the visual acuity of primates is highly developed. The location of their eyes gives them binocular, stereoscopic vision and, to some, an appreciation of color difference. This enhancement of visual perception is associated with improved function of the hands—for grasping, exploring, touching, and feeling objects. Like children, monkeys acquire added information by carrying objects to their mouths and biting them for taste and texture. Whether their brain can fully utilize the information garnered by the eyes and hands and can associate the new information with past events and experience remains speculative.

Unlike their improved visual perception, the olfactory sense of simian primates, great apes, and man has undergone various degrees of diminution and a corresponding regression in the olfactory lobes of the brain. As a result, most primates are microsmatic (i.e., have a reduced olfactory sensibility) and thus are characterized by a shortened muzzle or snout; the long, doglike snout of the baboon is an interesting exception.

The true method of locomotion of nonhuman primates, both in trees and on the ground, is quadrupedal. Their heads are so set on the vertebral column that they can comfortably

look straight ahead and can sit up with their knees bent and the soles of their feet flat on the ground. Although human babies cannot squat in quite the same way, their precocious ability to sit unaided with their heads up indicates that they will eventually stand upright.

Nonhuman primates have adopted a wide variety of loco-motor patterns. Some monkeys, especially the tree dwellers, use their long arms to swing by pendulum action on the limbs of trees (brachiation); their shorter hind limbs are used only for holding firm when at rest, for the initial propulsion, and for landing. Tarsiers and galagos progress either on the ground or in trees by leaps and by kicking like kangaroos with their elongated and highly specialized hind limbs and feet. This type of saltatory locomotion has freed their long-fingered, skillful hands for the manipulation of food.

The dominant component of brachiation is arm-swinging (Figure 20), that is, the body, suspended from above, is pro-pelled through space by means of a rapid alternating move-ment of the arms. Capitalizing on pendulum action, a brachi-ating animal can move about effectively with minimum ef-fort. Lacking tails and equipped with relatively short legs, the long-armed gibbons and siamangs are true brachiators whose front limbs are so disproportionately long as to make quadrupedal walking awkward. The long prehensile tail of some South American monkeys serves as a balancing mechan-ism during brachiation and as a fifth limb and hand during rest. On the ground, these monkeys have a quadrupedal gait and sometimes assume a bipedal stance.

Among the strictly arboreal monkeys, spider monkeys (*Ateles*) brachiate only occasionally and have been called semibrachiators. The term *semibrachiation* is not really satis-factory because it does not tell whether the performance is only half as good as full brachiation, is carried out for only

half the time, or neither. The prehensile tail of these animals, which enables them to move to the ends of branches and to feed under them, is an important factor in improving their access to more food.

The large lemurs, sifakas *(Propithecus)*, and *Indris* jump and leap through the trees semierect and often "walk" on the ground on their hind legs like gibbons. Among the terrestrial monkeys, Celebes black apes and Japanese macaques, both quadrupedal, walk easily on their hind legs when both hands are occupied or when they are trying to raise their level of vision (Figure 27). All the great apes, including the orangutan whose limbs are specialized primarily for moving about in trees, can and do walk bipedally. Much of the locomotor activity of the knuckle-walking chimpanzees and gorillas requires an erect posture because they often carry armloads of food while walking bipedally.

Since the feet of tarsiers, lemurs, and lorises are small in relation to the supports to which they cling, nature has provided an enlarged great toe (hallux) and elongated fourth and fifth toes for grasping and clinging. The second finger of pottos and lorises is so reduced that when their foot is outstretched, the hallux and fifth toe are set at about a 90° angle to each other. On the other hand, some arboreal South American monkeys and orangs support their bodies on hooked lateral toes, making little use of their vestigial great toe. Equipped with only a reduced hallux, marmosets and some lemurs use their sharp, clawlike nails to dig into tree bark when climbing. Most arboreal monkeys oppose their big toe toward their other toes, and the latter simultaneously bend toward the former. Thus, when the hallux is placed on one side of a branch, the other toes grip the opposite side somewhat like a hand.

Mostly arboreal in infancy, the gorilla later becomes al-

Fig. 27. A Japanese macaque (*Macaca fuscata*) walking bipedally.

most exclusively terrestrial, probably because his large size cannot be easily supported by branches. Hooton, however, believes that the gorilla becomes painfully and imperfectly adapted for a terrestrial gait only late in life. At any rate, gorillas have a large great toe, smaller lateral toes, and a better developed calcaneus than any other ape, and their toes remain opposable.

The heel bones of club-footed and flat-footed human adults and newborns have been compared with those of adult orangs, baboons, and gorillas. Since the orang is arboreal, his foot is adapted very differently from that of the terrestrial baboon and gorilla. Judging from its gross structure, the cal-

caneus of the human neonate resembles that of the orang, that is, it is similar to that of climbing animals. Being quadrupedal and flat-footed, terrestrial baboons and gorillas transfer their weight to the calcaneus differently from man; hence the shape of this bone is different from that of man. When the arches of the human foot become flat and the stress of weight on the calcaneus changes, this bone gradually reshapes itself to assume a form somewhat like that of baboons and gorillas.

The hands of other primates suggest interesting evolutionary trends and perhaps exemplify ancestral stages in the evolution of the human hand. The hand of the tree shrew, for example, resembles a foot whose thumb, though more divergent, is hardly distinct from the rest of the fingers, all of which have keeled nails or claws and expanded touch pads. The thumb of living lemurs is mobile but not opposable like that of man whereas the short, slender thumb of the baboon hand, which has shifted functional emphasis farther to the thumb side, is opposable.

Stationary food sources, like leaves and fruit, allow the hand and fingers to drift away from opposability and manipulative skills; mobile food sources, such as insects and small vertebrates, favor the retention of prehensility and the central focal closure of the finger pads. Thus we see that environment, particularly the availability of food, has had an extraordinary impact on the development of primate anatomy.

Some monkeys have refined communal living to such a degree that they have developed elaborate forms of integrated social activity, the best examples of which are found among the baboons and macaques. Field studies of langur, macaque, and baboon troops have shown specific and elaborate patterns of social behavior, which are assets only when members of the group are able to communicate. Since primate speech is simple, monkeys must have other more intricate and subtle

means of communication—howls, grunts, shrieks, chirps, and chatterings—which channel their meanings in ways not yet understood by man.

All animals have acquired at least basic patterns of communication whereby they signal such messages as greeting, mating, threat, danger, and flight. Nonhuman primates share these patterns with the other mammals, including man. For example, they signal a greeting by such facial expressions as grimacing, smacking, and protruding their lips; threat or attack by knitting their brow, opening their mouth, baring their teeth, and flattening their ears; excitement by grinning with or without frowning. These facial expressions are sometimes accompanied by growls, grunts, clicks, bleats, and whines and always by such postural attitudes as dancing, walking backwards, shaking branches, or pounding the chest. Each action, individually or in complex, has a specific meaning which sometimes varies with the species. For example, when angered, prosimians flatten their well-developed ears against their head and prepare to attack whereas baboons and macaques generally use this same gesture to convey appeasement. Lemurs often use their tail to indicate the mood of the moment, and monkeys as well as apes groom themselves to indicate social acceptance or contentment and present their buttocks to an aggressor to convey submission and willingness to cease hostilities. Females, however, use this last gesture to entice desirable males.

Certain primates, like baboons, for whom vocal utterances are subsidiary to gestures, depend more on the latter to mediate their social relations (Washburn and De Vore, 1961), perhaps because their social group is so compact that they are usually within sight of each other. When vocal sounds, such as a warning bark, are uttered without gestures to indicate danger, defiance, anger, fear, frustration, or contentment,

they serve mainly to draw attention to the animal which then gestures and postures. The importance of these posturings in communication can be judged from an experiment in which the vocalizations emitted by some primates during earlier stressful situations were recorded and played back to the same troop under quiet conditions without always eliciting the previous response. Schultz (1961) calls the cumulative expression of postures, gestures, and facial expressions by monkeys and apes their "intricate and voluminous *silent vocabulary*" during social intercourse.

The vocal expressions of these primates, as well as the anatomical and functional performance of their larynges, vary considerably. For example, the vocal folds of prosimians range from the nearly membranous structures of tarsiers to those of lemurs which have a substrate of muscles. Most monkeys have intralaryngeal outpocketings for resonance, but some even have extralaryngeal air sacs. The extremely long, sharp-edged vocal folds of capuchins and spider monkeys enable them to produce loud noises which are, however, almost sibilant compared with the prodigious roars produced by the long, sharp vocal folds of the howler which are supported by long, deep ventricles and reinforced by a great bony dilatation of the hyoid bone. A perforation in the epiglottis serves as a passage between the larynx and the membranous air sac which emerges from the middle of the hyoid bone of Old World monkeys and acts as a resonator. The cartilaginous skeleton of these larynges gives them plasticity. The pharynx is only loosely connected with the larynx, and the soft palate terminates in a uvula, not found in other mammals. Nonhuman primates use this apparatus to produce modulated and complex sounds unavailable to man.

Although the ape larynx appears to be superficially similar to that of man, it differs in many details, even among differ-

ent apes. An air sac is located extremely high against the base of the tongue, and the soft palate is almost inseparable from the epiglottis at the sides. In addition to the absence of a neurological base for speech in the great apes, their vocal apparatus has a limited ability to produce sounds.

The vocal expressions of primates are as varied as their forms and habits. Except for some lemurs, which have a complicated and expressive vocabulary, prosimians are generally silent, limiting their vocal utterances to calls of distress, mating, or the proclamation of territory. Some monkeys "chatter" frequently; for example, more than thirty expressions have been recorded for Japanese macaques; others bark, screech, cackle, or grunt. The whines and soft mumblings of the gibbon sometimes grow to astoundingly loud, almost metallic hoots that differ in pitch and rate of repetition. Siamangs emit such astounding vocal sounds that it is sometimes difficult to believe that they are being produced by animals. Howler monkeys use their notorious roar, often in chorus, to assert territorial rights, to convey alarm, and to indicate troop location and sudden changes in climatic conditions. Then again a male stretched comfortably on a limb sometimes produces deafening sounds for no apparent reason! Short on vocalization, orangs occasionally groan or roar like gorillas, which are usually content to grunt, murmur quietly, or say nothing at all. The loquacious chimpanzee expresses his emotions readily, often, and without restraint; some assiduous behaviorists have recognized more than thirty sounds, dubbed words, in their "vocabulary."

Speech as we know it, however, seems to be beyond the ken of nonhuman primates. Nevertheless, several investigators have tried to teach chimpanzees to use symbols in order to express their desires and to communicate their "thoughts." For example, in a dedicated, if somewhat anthropomorphic,

study of Sarah, a young chimpanzee, Premack and Premack (1972) made endless observations of her ability to learn and to communicate symbolically by means of different-shaped blocks and cutouts. Assuming that language is a form of general system communication, of which human speech is a refined form, the authors hypothesized that those features of human language that are usually considered to be uniquely human belong to a more generalized system and can therefore be distinguished from those that are part of the gestalt of human information. They argued that if an ape could be taught the rudiments of human language it would be possible to separate or otherwise clarify the underlying system common to the two and single out the one which is specifically human.

Rather than recite in detail their experimental approach and that of other authors who have done similar studies, I will mention only their interesting, if somewhat specious, conclusion. When subjected to a basic, simple training program, whose aim is to reduce complex notions to easily "learnable" language, apes (i.e., young chimpanzees) can learn a simple language code that includes the natural features of (human?) language. In defending this conclusion, the authors admit that apes do not perform as well and as quickly as human adults, but more like two-year-old children. Anticipating the skeptical reactions of some of their readers, Premack and Premack (1972) defended their position by dismissing as highly prejudiced in favor of their own species those scientists who are impressed only when apes perform prodigiously but at the same time deny that apes have human capabilities. This kind of defensiveness is unbecoming in scientific work and does little to strengthen the authors' hope that their findings will dispel prejudice and lead to other attempts to teach language to nonhuman animals.

No account of language and speech would be complete without some reference to the observations of Sherwood Washburn (1968). He points out that language, speech in particular, is wholly cerebral whereas nonverbal communication, such as the expression of emotion, can be achieved without much involvement of the brain. Language is directly related to large specific areas of the brain, indirectly to several others. Because language depends to a large extent upon memory, it presupposes "foresight and planning." Washburn reminds us that the evolution of language is intimately linked with the "behaviors it makes possible." Language conveys specific information about the environment by giving names to objects, moods, and circumstances, a uniquely human act; however, nonhuman communication uses sounds (such as the warning cries that distinguish arboreal from terrestrial predators) only to emphasize expression, gestural postures, and other actions which communicate emotional states or social intent. With one cry, the monkeys take to ground; with another, they climb trees; but in neither case do they know what predator they are escaping from. In human language or speech, sounds have specific and mostly precise meaning, and it is this ability to assign names to objects and to specific emotional and social moods that makes human speech and language unique. Washburn suggests that human language developed first because of tools, which, in turn, focused interest on the environment and the communication of many kinds of information. This aspect of evolution, says Washburn, had a greater impact on the development of human society and culture than any other event because to name things requires recognition, learning, and memory, the essence of speech. Apes and monkeys have been conditioned to respond to sounds or words but they cannot learn to talk. Man, on the other hand, easily learns to speak one or several languages. Nevertheless, speech

seems to be a relatively recent phenomenon. Having demonstrated that monkeys and apes cannot name things and therefore do not have even the rudiments of language, Washburn concludes that language represents a complete break with the past. Man, therefore, might well be defined as an animal who talks. In that sense, no other primate is an adequate substitute for man in experimental research.

But what about the ability of nonhuman primates to perceive the communications conveyed by speech or other sounds? In other words, what about nonprimate hearing? Considerable, if somewhat fragmented, data suggest that chimpanzees can hear up to 30,000 Hz and that the upper limits of several monkeys tested are above those of human range. Such prosimians as lemurs and galagos have superlative hearing acuity. Both in them and in New World monkeys, the floor of the middle ear cavity balloons out to form a structure called the tympanic bulla. In tree shrews and lemurs only, this structure encloses the tympanic ring which in lorises and New World monkeys is exposed on the side of the skull. Both tarsiers and Old World families have a bony external auditory meatus which is lacking or much shortened in the Lorisidae and New World families.

Man, the highest of the primates, is at the low end of the frequency scale; his sensitivity to high frequencies is the first to go. Crab-eating and pig-tailed macaques, on the other hand, have an upper sensitivity of about 45,000 Hz. All of the other anthropoids tested range between these two extremes. Among the more primitive primates, sensitivity to sound gradually increases at the upper range; for example, ring-tailed lemurs hear up to 75,000 Hz, tree shrews 90,000 Hz. Man, however, possesses the best low-frequency sensitivity and discrimination. The extraordinarily large ears of such nocturnal prosimians as tarsiers and galagos can be oriented toward

special or startling sounds without any movement of the head in the same direction. When alert, their ears are erect and fully extended. The edge of tarsier ears is constantly undulating, leading one to surmise that the external ears of these animals, like those of bats, are best adapted to receive high-frequency sounds.

The observations and topics included in this brief summary of the natural history of primates have been highly selective and no doubt reflect my own bias. In the next chapters we will explore the life-style of nonhuman primates and in so doing discover that it is what they do, even more than what they are, that provides the rationale for classifying them with man.

CHAPTER 3

PRIMATE BEHAVIOR

Whether studies of all aspects of primate behavior contribute to our knowledge of human behavior is open to question, but certainly those on primate sociality indicate that the patterns of social and sexual behavior of some primates bear some resemblance to those of man. This, however, must not be taken to imply that the behavior of nonhuman primates is a pale reflection of man's. On the contrary, those who have studied it seriously find it highly complex in its own right. Thelma Rowell (1972) suggests that the advantage of studying the behavior of nonhuman primates rather than that of man is not that the former is simpler but that these animals have a shorter life-span than man and hence longitudinal studies of their behavior can encompass more stages. Moreover, in watching animals interact, the observer is not so distracted by what they are doing and how they rationalize as he is likely to be with people but can concentrate dispassionately on the event being observed. Nevertheless, because their

NOTE: This chapter is based on a lecture in the Wesley W. Spink Lectures on Comparative Medicine presented at the University of Minnesota, Duluth campus, on October 15, 1975.

overt behavior so often resembles that of man, one may mistakenly assume a similar motivation for the behavior of non-human primates. The strongest motive for studying nonhuman primate behavior is the belief that it can elucidate the stages through which human behavior evolved. Underlying all such studies is the hope that we can thereby gain a better understanding of ourselves. Current research, which correlates certain social structures with the quality of the environment in which they developed, will be discussed below.

Except for the nocturnal prosimians and the New World owl monkey, most primates live in groups, which vary in size according to the life-style and habitat of each species. Sodality, of course, is not peculiar to primates, but in some of the more terrestrial forms of the order the social hierarchy is so highly structured and complex that it has attracted much attention. Eisenberg (1966), for example, believes that social grouping among primates is much more complicated than the mother-infant grouping among many other mammals. Unlike the diurnal, mobile inhabitants of the grasslands which form social groups of various sizes and composition, the nocturnal, somewhat sedentary, forest-dwelling prosimians are mostly solitary, that is, spread out over a wide territory in search of food, but nonetheless aware of their neighbors. The same patterns can be observed among other mammals, too; nocturnal carnivores, for example, are mostly solitary whereas diurnals, like most canids and some cats (e.g., lions), are gregarious. Evidently, the former must rely chiefly on olfactory, auditory, and perhaps some tactile cues for intraspecific communication; the latter are endowed with superior visual and auditory acuity (Rowell, 1972). Nearly all nocturnal, as well as many diurnal, mammals have some type of scent gland, and all of them have well-developed vibrissae which help them to orient to the environment they cannot see during the

night. The functional adaptation of each of these mechanisms is self-evident.

Currently, studies in chemical communication among gregarious animals are extremely popular. When highly developed, this type of communication is the most efficient of all, since only a few molecules of certain substances can convey a message to a conspecific member some distance away without revealing the presence of either animal to possible predators. Much of what is known about the prevalence, composition, and effect of pheromones among primates has been pioneered by Michael and his colleagues who since 1960 have done most of their research with rhesus monkeys and have now extended their studies to human beings. (Michael and Keverne, 1968, 1970). They have amassed a sizable amount of data, which will be briefly summarized here. This group has proposed that around the onset of ovulation a chemical substance in the vaginal fluid of rhesus monkeys signals to the male the female's physiological condition and her predisposition to copulate. This substance, called (perhaps with tongue in cheek) copulin, is apparently composed of six short-chain aliphatic acids: acetic, butyric, isobutyric, propionic, isovaleric, and isocaproic. When the "correct" mixture of these fatty acids was smeared on the perineum of ovariectomized female rhesus, which are not normally attractive to males, the males became concupiscent. Later, Michael states that he had isolated the same fatty acids from the vaginal secretions of women. A carefully formulated decoction of these substances is now being investigated with volunteer wives as experimental subjects. These women apply some of the substance to their chests in the evening for their bed partners to sniff; they then prepare a report on their sexual activity with and without the aid of the pheromone. Despite Dr. Michael's reports, it is still de-

batable whether monkeys or human beings actually secrete pheromones. Since, as noted above, man is microsmatic (i.e., has a reduced olfactory sensibility), the many subtle odors in his environment drift in and out of his consciousness and are seldom explicit to him. This is in contrast to the experience of such macrosmatic animals as rodents whose olfactory discrimination reaches the level of molecular differentiation. The specificity and accuracy of the animal's response are based, among others, on his selective reactions to a particular signal, which he can distinguish from a background of numerous other signals.

Evidence of individual and species-specific odors has been established by the ability of some animals (especially mice) to discriminate among members of their own or other species. Human beings have various biological mechanisms which impart to them characteristic odors. We are all well equipped with sebaceous and apocrine glands, both of which secrete substances with an odor that is characteristic of the species as well as of the individual. The same is true of vaginal secretions and of the smegma that accumulates under the penile prepuce. Characteristic odors are also produced secondarily by the action of microflora on these secretions. These colonies of microorganisms differ from individual to individual and from one specific locus to another on any one animal's topography. Like that of many other mammals, the urine of human males contains several musky, odorous substances, among them androstan, to which they are generally insensitive but of which women are keenly aware during some phases of the ovarian cycle. Thus certain types of odors emitted by mammals elicit various kinds of responses, some of them sex-related but none of them fully understood. According to abundant evidence, sensitivity to these odors is modulated by

the neuroendocrine system. Some data indicate that dogs, billy goats, and bulls are attracted to certain body odors of women but not to those of men.

Culturally, human beings no longer respond favorably to natural odors. In fact, together with antiperspirants, an endless array of substances has been produced to combat or minimize odors emanating from the axilla, vagina, and total body. Yet at the same time a colossal industry has been built around the manufacture of perfumes, which are, in fact, exogenous pheromones, for both men and women. Ironically enough, the bases of some of the most expensive perfumes are the pheromone-producing substances of such animals as the musk ox and civet cat.

Michael and his group seem to have misinterpreted or perhaps oversimplified the biological function of pheromones in nonhuman and human primates by minimizing or ignoring the impact of their respective cultures on the behavioral responses of man and other animals to odors. They fail to take into account individual differences and past experience. Libido in human and simian primates is a highly complex phenomenon, controlled not only by the endocrine system and the individual's social history, but also by the mysterious, unknown regions of the cerebral cortex and can, therefore, scarcely depend to any great extent on olfactory cues, which more often than not are subliminal. No one except Michael's own group seems to have successfully repeated his observations. In fact, some well-designed, carefully controlled experiments at the Caribbean Primate Center and the Wisconsin Regional Primate Research Center have consistently failed to replicate Michael's results. In a recent review* Dr. Robert Goy, director of the Wisconsin Regional Primate Research

---

Medical World News (1974), "Smells: Surer than sounds or sights." September, 36C-36D.

Center, has made some keen observations on primate pheromones and behavior. Goy doubts that functional pheromones
reinforce the libidinous behavior of either man or nonhuman
primates and believes that they are at best accessory or redundant cues to the cadre of their sexual behavior.

In a series of experiments with adult male and spayed female rhesus monkeys, Goldfoot et al. (1975) showed that vaginal lavages obtained from estrogen-treated donor females
had no significant sexual stimulatory properties when applied
to spayed nonestrogenized recipient females. However, when
paired with estrogenized females, all but one of the males
copulated and ejaculated, and some of the males showed
moderate increases in sexual behavior only when the vaginal
lavage tested was contaminated with twenty-four-hour-old
ejaculate. When purified aliphatic acids were applied to
spayed nonestrogenized recipients, only slight changes were
observed in the sexual behavior of the males.

When these authors attempted to quantify the short-chain
aliphatic acids, the alleged active component of pheromones
in vaginal lavages, they found no detectable levels in spayed
females. But daily treatment with estradiol benzoate for six
to ten days (25 μg/day i.m.) induced measurable levels of acetic, propionic, and butyric acids, and exposures for six months
produced a fairly constant concentration of aliphatic acid.
The addition of male ejaculate, however, caused a fivefold
elevation in the levels of aliphatic acids in the vaginal con-
tents.

A study of some intact females throughout an entire men-
strual cycle showed that the peak values of aliphatic acids occurred during the luteal phase, several days after the presumed
ovulation. For the majority of the males studied, the application of vaginal lavages from midcycle donors did not significantly increase their copulation with spayed, nonestrogenized

recipient females. Moreover, the aliphatic acid determinations suggest that facility in copulation is not always associated with increased concentrations of these substances since the largest increases in aliphatic acids were found after copulation and ejaculation. Goldfoot and his colleagues (1975), when the results of their studies done during the luteal phase in cycles free of copulation and after progesterone treatment of spayed estrogenized females were compared with those from Michael's laboratory, concluded that any positive effects observed probably depended on either associative learning or the extinction and disinhibition of sexual interest.

Like human beings, simian primates are greatly preoccupied with sex. Since their libido is less influenced by seasonal changes than that of many other mammals, they frequently engage in sexual activities, sometimes without regard to the time of the ovulatory cycle. Thus, investigations into the sexual behavior of nonhuman primates are clearly pertinent to human sexuality, especially when they focus on the individuality of the experimental subjects and record their past experiences, hormonal milieu, and the effect of surgical and hormonal intervention and replacement hormonal therapy in castrated animals.

Beach (1947) first suggested that sexual behavior is not as dependent on gonadal hormones in nonhuman primates as it is in other mammals, and in man is the most free of hormonal control (Ford and Beach, 1951). Beach and other authors have even shown that in several mammals some aspects of sexual behavior persist after castration, when no residues of testicular androgens could be detected.

The sexual responses of castrated men vary from none to frequent. To shed some light on this subject, Phoenix and his colleagues (1973a, 1974; Resko, 1972) at the Oregon Re-

gional Primate Research Center first studied intact animals, then castrated them, and finally treated them with hormones. They also studied the sexual behavior of experimentally produced female pseudohermaphrodites.

In normal, intact male rhesus monkeys, these authors found no exact correlation between the levels of plasma testosterone and the amount and quality of sexual performances. Furthermore, the behavior of each animal was unique and, once known, largely predictable. Castration had little effect on the sexual behavior of some males but in others the behavior was either greatly reduced or completely suppressed. Thirteen weeks after castration, very few of the animals had residual plasma testosterone levels. The 186-613 pg/ml of androgen found in some animals one year after castration probably originated from the adrenal cortex. As in intact animals, the amount of residual circulating androgens in castrated animals had no relation to the frequency of sexual performances, regardless of their source. However, exogenous testosterone administered to these castrated animals restored their sexual performance to normal precastration levels. Thus individual animals respond differently to the same amounts of hormones; moreover, the traces of androgens in castrated animals may maintain sexual behavior in some but not all animals. It is also possible that the residual sex behavior is not endocrine-related.

When plasma testosterone and luteinizing hormone were measured in intact and vasectomized adult rhesus monkeys, the results were of special interest. These hormones were quantified from the peripheral vein blood drawn before and 50, 80, and 140 minutes after coition. In both groups, the amounts of testosterone and luteinizing hormone did not differ before and after ejaculation. The same results were re-

corded for another group of male rhesus monkeys before and after electroejaculation and concur with those obtained in a limited number of similar studies in man.

Sexual behaviorists in the past overemphasized the role of the male in sexual behavior and generally overstressed the passive role of the female. Observations by Beach and his co-workers (1969) on dogs and by several others on human beings present a different picture. The studies in Dr. Phoenix's laboratory (Johnson and Phoenix, 1975) with rhesus monkeys leave no doubt that females, though generally more subtle than males, largely control the sex behavior of their male counterparts. In groups of adult rhesus monkeys, the females were permitted to control the occurrence of paired sex tests. In these studies that employed ovariectomized animals, some of which were treated with estradiol, others with two dosages of testosterone, and still others with estradiol plus testosterone or estradiol plus dexamethasone, or not treated at all, the authors observed how various treatments affected female attractiveness (the female as an effective sexual stimulus), proceptivity (the extent to which the female seeks out the male and elicits sexual behavior), and receptivity (willingness to copulate). They discovered that both estrogen and androgen affected the sexual attractiveness of females, estrogen alone enhancing it, but the effect of androgen depended on the dose. Both estrogen and androgen increased female proceptivity, whereas estradiol but not testosterone somewhat stimulated receptivity. Thus the commonly accepted hypothesis that testosterone is the "libidinal hormone" in female primates was not supported by these workers.

The conviction that some patterns of sexual behavior are not entirely dependent on endocrine is also supported by Eaton and his colleagues' (1973) observations on experimentally produced rhesus female pseudohermaphrodites. Pseudoher-

maphrodites can be produced by injecting pregnant animals with 5 to 25 mg testosterone propionate daily throughout the middle trimester of the normal 168-day gestation. All female infants born after this treatment were pseudohermaphrodites with normal ovaries, uterus, and oviduct but also with a phallus, an empty scrotum, and no external vaginal opening. The young were weaned at the usual 90 to 110 days and randomly assigned to social test groups consisting of five or six normal males and females of the same age together with one or more prepubertally castrated males. Longitudinal studies of the behavior of the pseudohermaphrodites for four years show that the frequency with which these pseudohermaphrodites displayed juvenile patterns of threat and aggression, play-initiation, rough-and-tumble and chase was intermediate between that of normal males and females. The development of the mounting pattern resembled that of control males.

When the adult pseudohermaphroditic females were subsequently ovariectomized, treated with exogenous testosterone, and placed with formerly normal but ovariectomized test females which had been primed with estrogen, they were threatened and bitten by the latter. However, immediately after the castrated pseudohermaphroditic females had been treated with testosterone, they began to display more pronounced male sex behavior than formerly normal ovariectomized control females similarly treated. After thirty weeks of treatment, in fact, one pseudohermaphroditic female with a penis as large as that of a normal male achieved intromission and ejaculated with a test female.

An analysis of the sexual and social behavior of experimentally produced adult female psuedohermaphrodites indicates that they are singularly sensitive to testosterone propionate. Their aggressiveness before treatment, like their earlier behavior as juveniles, reflects the prenatal masculinizing effect of

exogenous androgen on the brain of developing fetuses. Exogenous androgen, then, can modify the central nervous system of genotypic female fetuses in such a way as to dispose them to acquire predominantly masculine patterns of behavior.

Now that laboratories are breeding their own experimental animals, we have increased our stock of information and profited from observations on their sexual behavior. Breeding animals in the laboratory requires the careful selection of compatible pairs. When rhesus monkeys at the Oregon primate center are bred for the timed pregnancies needed in perinatal studies, the males, large and often ferocious, are kept in separate cages where they must not be stressed or disturbed; the females are brought to them in separate, small holding cages. Whether the male accepts or rejects the female offered to him depends on many factors, about which little is known. What signals the female gives the male during the interchange is likewise not known, but she is not introduced into the male's cage until he registers his overt approval, which, who knows, may be conveyed by her. To place her with the male before approval is to court hostility and attacks.

In some species of nonhuman primates, the sex skin and genitalia of females become swollen near the time of ovulation. For example, the buttocks of pigtail macaques (*Macaca nemestrina*) and of Celebes black apes (*M. nigra*) grow to disproportionate sizes (Figure 28) and turn a brilliant red in response to endogenous or exogenous estrogenic stimulation. The vulva of most primates secretes viscid substances around the time of ovulation and we have already discussed whether or not they attract the attention of the males (see the discussion of pheromones earlier in this chapter). Since the vagina of nonhuman primates, like that of women, is devoid of glands, this material is probably the combined secretion of

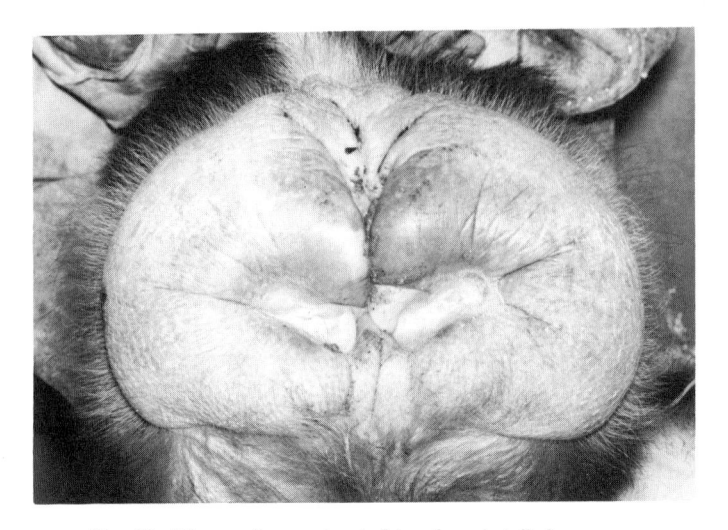

Fig. 28. The swollen perineal skin of a pig-tailed macaque
(*Macaca nemestrina*) at the time of ovulation.

the uterus and cervical glands and the greater vestibular and labial glands. Female lemurs accept males only during or just before ovulation; outside that period, males are indifferent or even aggressive to females, and vice versa. When vaginal lavages of ovulating female lemurs are smeared on the perineum of nonovulating animals, the sexual attraction of males is apparently aroused. Male chimpanzees rarely make sexual advances until the females are midway through the ovarian cycle (when their genitalia become greatly swollen) and actively seek the attention of males. In this respect macaque behavior is more nearly akin to that of man than of the chimpanzee.

After this rather lengthy digression on the role of chemical communication and the various expressions of sexual behavior among primates, let us return to the life-style of primates in general. Napier (1962) tells us that about 90 percent of ex-

tant nonhuman primates live in trees; the others inhabit sa-
vannahs, steppes, and deserts, and some share their territory
with man. Whatever their environment most terrestrial primates except adult gorillas are adept climbers and still make
good use of trees. The majority of these animals inhabit tropical and subtropical zones, but the range of some macaques
and langurs extends to very cold regions such as Japan and
the Himalayas. Dolinow (1971) emphasizes that environment
deeply affects the social structure of terrestrial species living
in temperate zones. She points out that studies of rhesus monkeys living in the cities and forests of India clearly show these
troops behaving differently in the two environments. In general, animals that live in tropical forests where the food supply is constant throughout the years and predators are few
live in relatively small societies and move over a limited territory. Fully or partially terrestrial animals which live in areas
with few or no trees, however, have had to change their life-
style. Because food is invariably scarce, they need a greater
range and consequently have developed a social structure that
provides adequate protection against predators. Animals like
the Colobinae, which have a specialized diet, must sometimes
travel great distances to the source of their food supply.

The environment has had a profound effect not only on
the sociality of primates but on their physiological adapta-
tions to it as well. Like all other animals that live in arid environments, primates inhabiting steppes and deserts have developed water-conserving mechanisms. Evidence of workers
who have studied the troop movements of howler and spider
monkeys in the rain forests of Central and South America indicates that the abundance of food there is nearly inexhaust-
ible and that as a result the troops move about only for exer-
cise and territorial survey. In contrast, some species of baboons must cover large stretches of open, dessicated grass-

lands to obtain enough food (Simonds, 1974) and must, in the process, expose themselves to predators. It is these animals, rather than those who live in the relative comfort of the forest, that show the greatest adaptability and variability in social structures. Their group unity is more complex and precise, and the groups are larger; in addition, the scarcity of food has forced them to be less selective. In this connection, I recall the reaction of a troop of baboons to a grass fire in the Serengeti a few years ago. Forming a line some thirty feet away from the fire, the animals caught and ate handfuls of insects flushed by the smoke and heat. One even devoured a small mammal!

Although most simian primates, including baboons and chimpanzees, are mainly vegetarian, they are occasionally omnivorous; under certain circumstances, some have even been known to develop a predilection for meat. The consumption of meat by these animals, however, may be restricted to certain populations. For an excellent study of how predation and meat consumption were incorporated into the tradition of a group of baboons, I recommend the meticulous reports of Strum (1975a, b), who collected her data during a series of field studies. While observing the social behavior of a troop of olive baboons (*Papio cyanocephalus*) in Kenya, the author, as Harding (1973) had done before her, recorded a very high rate of kills and meat consumption among them. In the beginning and during the early months of the study, small animals, such as hares, birds, and even young antelope, were killed and eaten by the adult males and a few females. But as time passed, fewer ungulates and more hares and birds were hunted. Strum associated this change with significant changes in participation both in hunting and in meat-getting. At first, the adult males did practically all the killing and eating; only occasionally did the adult females take part in the killing and

rarely in the eating. Later, however, females and juveniles engaged in the killing more and more frequently; with their participation came changes in the types of prey and the number of species killed.

Juvenile participation transformed the kills into focal points of social interaction since the mothers frequently joined their young in these activities. Juvenile males and play groups began by approaching and investigating the kills (mostly birds and hares), progressed to capture, and finally to consumption. By the end of the study, they had successfully captured a young Thomson's gazelle. In time, even the infants participated in the meat-eating. As they grew and gained some independence as members of play groups, they began to attend kills. Infants with the greatest predilection for eating meat were those who had special relationships with adult killer males.

Jane van Lawick-Goodall (1971) also observed both predation and the eating of meat by chimpanzees, who seemed to prefer killing other primates; on two occasions, they presumably ate human infants.

Teleki (1973) summarizes the predation and carnivorousness of chimpanzees by stressing the importance of predatory behavior as a demonstration of cooperation in the hunt and the sharing of the kill. This, however, is not unique to nonhuman primates or to man but is shared by many carnivores.

Although most of these observations on predation are anecdotal, they give us some insight into the origin and development of social traditions and culture among primate groups. Not only have nonhuman primates learned to kill and consume other animals, but they have done so without the use of tools or language.

Consumption, of course, is not the only or even the principal purpose of predation and kill. The baboon troops which

live in the open country of East Africa form organized groups for protection against such predators as leopards and cheetahs. To ensure maximal protection, they have achieved the cooperation of the males—large, powerful, aggressive—a cooperation which appears to be stronger than that found among other primate organizations. It must be emphasized that adaptation to a particular environment can be achieved by various kinds of social organizations and that even among individual species the ability to adjust to different environments varies. There is no doubt, however, that macaques and their near-relatives the baboons have attained the highest adaptability. The various cultures developed by these animals, whether or not within the same species, are directly related to their environment. The difficulty of bolstering such statements with data on feral animals is that in studying troops of wild animals one can never be absolutely sure that he has always identified each member of the troop correctly or that the animals will be on hand when he is ready to observe them. Only an accumulation of data on each animal can provide the insight needed to construct the dynamics of social structure. The most reliable data are those that have been gathered from observations of troops of animals whose territory is large enough to ensure them freedom of action and small enough to enable the observer to obtain a maximum amount of information on each member of the troop. Among the most competent scientists to observe the sociality of wild troops of primates are Washburn and De Vore (1961), Jay (1962), Schaller (1963), Carpenter (1964), and Eisenberg (1966). In the chapter that follows we will consider what can be learned from studies of the same animals over an extended period of time.

CHAPTER 4

SOCIAL BEHAVIOR
OF JAPANESE MACAQUES

W e were fortunate when in 1965 a natural troop of Japanese macaques (*Macaca fuscata*) was transported intact from Japan to the Oregon Regional Primate Research Center and housed in a two-acre grassy corral where the actions of every animal could be observed and recorded daily from two high towers. These rugged, mostly ground-dwelling monkeys can live so far north that some inhabit the snowy mountainous regions of Japan where they develop a rich protective fur and a thick layer of body fat. They fare well in Japan, where only man and some dogs are their enemies, and have thrived in Oregon even during the occasional periods of severe winter weather. They are so adaptable that they are totally omnivorous and eat and digest nearly anything, whether edible or semi-edible. The Japanese ethologists who have studied them have concluded that they live in close-knit social groups, with individually characteristic but somewhat similar social orders. In studies of wild troops since 1952, Japanese

NOTE: This chapter is based on a lecture in the Wesley W. Spink Lectures on Comparative Medicine presented at the University of Minnesota, Duluth campus, on October 15, 1975.

scientists have observed that the basic multimale-multifemale organization varies so that under certain conditions some all-male and all-female groups are formed.

Because of the sheer volume of studies of primate behavior, an overview is regrettably out of the question here. I shall, therefore, limit my discussion to one study, selected because of its thoroughness, intimacy, and pertinence; it is also the one I am most familiar with. The data were gathered at the Oregon Regional Primate Research Center and are a distillation of Gray Eaton's observations on the troop of Japanese macaques mentioned above. Eaton and his associates have spent several years continuously studying the social order of this typical multimale-multifemale, once-feral troop of Japanese macaques. Apparently, these monkeys had roamed free near Mihara City in the Hiroshima prefecture until complaints from local farmers prompted the Japan Monkey Centre to capture the entire troop by providing them with an open corral where they all came to eat regularly. When they were finally caught—forty-six of the original troop of forty-nine—the Japan center shipped them as a gift to the Oregon center. The account that follows is based on Eaton's (1975) observations of these animals.

Eaton, who is convinced that the partial confinement of the corral has had relatively little effect on the social order of the troop, believes that the two basic and related characteristics of the social order of Japanese macaques are a dominance hierarchy and the social roles of various classes of individuals within the troops. The former is so evident that it commands more attention than the latter. The troop has a high degree of aggression which develops centrifugally from the dominant males in the center and which is observed not only in some old females who attack other females with impunity but also in other females who threaten and chase adult peripheral

males. When the behavior of all the individuals within the group is analyzed, however, it is clear that the social roles played by these individuals are more important than dominance in determining the structure of the social order. These roles constitute the essence of a society as opposed to a simple aggregation of individuals. Because of the function of each individual in the group, the members learn to adapt to the environment in many ways that would be impossible for individuals. The roles can be as simple and interchangeable as watching for predators or as complex and fixed as rearing infants. Therefore, the importance of the dominance hierarchy in determining the structure of the social order can be measured not so much by dominance per se as by the differentiation of individual roles within the hierarchy.

The top rank is usually occupied by an adult male who sometimes does not attain this position until he is eighteen or nineteen years old (males normally reach puberty at four and full body growth at eight to ten years of age). Immediately below this leader (Figure 29) or "alpha" male are, typically, five or six "subleader" males. Next, most of the adult females, who reach puberty at three and full body growth at six to eight years of age, form the middle of the hierarchy along with their infant and juvenile offspring. All other adult males are located at the bottom or around the periphery of the hierarchy.

Dominance rank, which is determined by observing the results of aggressive encounters, is basically linear, but occasionally this is reversed. For example, animal A chases animal B, and B chases C, but C chases A. Hierarchy, however, cannot be computed simply by counting the frequency of aggressive acts. Originally, the number of times a presumably high-ranking animal is attacked by a lower ranking animal determines the linear order. For example, the alpha male is not at-

tacked, and the second-ranking male is attacked only by Alpha I. The third-ranking male is attacked only by Alpha I and the second-ranking member, and so on. In this way, high rank is not necessarily correlated with a high frequency of aggressive behavior. Actually, the second- and third-ranking animals are both more aggressive than Alpha I because, among other reasons, Alpha I receives more "respect" from the other troop members and does not need to threaten or attack other animals to maintain his position. When the troop is fed, Alpha is given more personal space than the second in command so that when a choice morsel is thrown near either of them, the second-ranking male does a great deal of chasing whereas Alpha I's presence alone is enough to keep the others away.

The low but positive correlation between the frequency of aggression and dominance rank is related to the different social roles played by the alpha and subleader males. The alpha male only occasionally puts a stop to fights among troop members; it is the subleader males who chase the more aggressive combatants (Figure 30) and keep order according to specific codes. Attacked animals frequently run toward a subleader male, who then chases the pursuing monkey for having inadvertently directed threatening gestures toward him. Yet, in a fight, subleader males attack whoever is on top. The losing monkey usually cringes, screams, and grimaces (all signs of submission that tend to minimize attack) whereas the other monkey growls and gapes and flattens his ears (threatening or aggressive gestures that are likely to induce attack) (Figures 31, 32, 33). When the alpha male occasionally threatens the combatants, this seems to indicate to the subleaders that they may attack and stop the fight. Sometimes, when these males are uncertain about what to do, they alternately threaten the combatants and turn to look at the alpha male.

The latter sometimes looks away as if ignoring the affair, sometimes threatens the subleader male, or joins him in threatening the combatants. In the first case, the subleader male may or may not attack; in the second, he leaves the area; and in the third, he attacks and disperses the combatants.

Eaton has observed only one male in the Oregon troop rise from the peripheral to the subleader class. The animal did this very quickly after the number two male attacked and wounded number three. Then the number four male, the highest ranking of the peripheral males, independently attacked and defeated number three. Thus he rose from the peripheral to the subleader class, and the number three male withdrew for several months but eventually fought his way back to the lowest post of the subleader males by chasing and threatening peripheral males. The ascendancy of the once peripheral male was so rapid that he was not immediately adroit in the subleader role of policing. He assumed the responsibility but was often ineffective in carrying it out. During some fights, he became so excited that he attacked animals that were not even involved. Eventually, however, he learned his role and became one of the best in quieting aggression.

The main roles of the alpha male appear to be to direct the movement of the troop and to defend it. In the enclosure of the Oregon Regional Primate Research Center, where troop movements are, of course, limited, his first role is necessarily curtailed; however, he is a stout defender. This role is especially important when the troop is rounded up once a year for husbandry purposes. To capture the animals, the handlers detach small groups from the main troop and herd them toward the catching pen. During these maneuvers, the leader, disregarding his own safety, leaves the troop and joins the small herded group. But, unlike the others, he faces the handlers and continuously threatens them, backing toward the

catching pen. When the small group is eventually driven into the pen, the leader dashes through the line of animal handlers and returns to the main troop where he continues his "defense" tactics.

When an occasional sick or wounded animal has to be removed from the enclosure, the alpha male again assumes the role of leader and heads a group of low-ranking males, some subleader males, and some adult females who charge and threaten the "intruders." The other subleader males move away from the alpha male and his cohorts and stay close to the mothers and infants in the main troop. The threatening animals come very close to the handlers but do not actually attack. The strongest stimulus to elicit threats is the cry of an infant; then the alpha male and the other adult males, shoulder to shoulder, advance toward the intruders, growling, bobbing their heads, and slapping the ground with their hands. Oddly enough, however, these adults make no attempt to prevent the handlers from retrieving a dead monkey unless it is an infant. A mother carries her dead infant for three or four days and then abandons it.

Adult female animals raise and protect their offspring with alacrity and adroitness. Since newborn infants are able to cling only to their mothers' breast and abdomen, they are supported by their mothers with one hand during this early period. Soon, however, the infants attempt to crawl away from their mothers but are held back, sometimes with one foot if the mother's hands are occupied with food or otherwise engaged. When the infants are about two weeks old, the mothers repeatedly place them on the ground and then back away and smack their lips, coaxing the young to stagger toward them. At the same time, the infants begin to ride piggyback on their mothers (Figure 34). These developmental stages are, of course, characterized by individual differences.

Fig. 29. Alpha I, the dominant male in the colony of Japanese macaques. He is old and has lost his right eye.

Fig. 30. Aggressive behavior in Janese macaques.

Fig. 31. Two female Japanese macaques fighting. (Courtesy of Kurt Modahl, Oregon Regional Primate Research Center.)

Fig. 32. The second-ranking m coming over to break up the f (Courtesy of Kurt Modahl, Ore Regional Primate Research Cen

Fig. 33. The second-ranking male chasing one of the females away. (Courtesy of Kurt Modahl, Oregon Regional Primate Research Center.)

Fig. 34. A female Japanese macaque carrying her month-old child piggyback.

Fig. 35. A swaggering male Japanese macaque.

Some mothers encourage their infants to ride on their backs, others do not. Infants are always a source of intense interest to other females, particularly their older sisters and adult females who do not have infants of their own at the time. But no female is permitted by the mother to pick up her infant until it is several weeks old. Occasionally, an adroit female manages to touch an infant by grooming and distracting the mother. Such jealous protection of infant Japanese macaques is in direct contrast with Jay's (1962) observations on Indian langurs. Apparently only hours after birth the mother passes her infant to other adult females. One wonders about the differences in temperament and personality that emerge from such divergent early methods of maternal care.

As the infants venture farther and farther away from their mothers and mature into juveniles, sex differences in behavior become apparent. Juvenile males spend much of their time in rough play with their peers whereas juvenile females spend their time grooming with their mothers and sisters. An occasional female will join a male play group but she does not stay long. Juvenile males seldom groom other monkeys, but they engage in self-grooming more frequently than females. All juveniles, both male and female, prefer to groom adults of their own sex rather than their peers.

Japanese macaque mothers play a major role in determining the eventual position of their offspring in the dominance hierarchy. In studies of wild troops, Japanese ethologists observed that an individual's rank is largely influenced by its mother's rank. How this is achieved can be observed in the Oregon animals during juvenile fights which occur mostly because one juvenile is playing too roughly. The injured juvenile screams and is rescued by its mother, who bites the offending juvenile; he, in turn, screams and is joined by his mother. The two mothers fight for a while, the dominant female and

her offspring eventually chasing the lower ranking female and her son or daughter from the scene of combat. After this has been repeated several times, the offspring of the lower ranking female runs away from that of the higher ranking female even when she is not there. Sometimes, however, unusually aggressive young males or females can rise above their mother's own rank on their own. Withal, however, most offspring remain equal to but not higher than their mother's rank.

How the mother's rank affects her son's rank among the peripheral males is uncertain. In feral troops, only the young sons of high-ranking females are allowed in the center of the troop; the other male offspring are driven to the periphery shortly after they have reached puberty or when they are about five years old. These ousted young may even have to fight for a place in the peripheral hierarchy. But Eaton observes that the mother's influence may still carry over because macaque fights are often "bluffing matches," and the sons of high-ranking females would presumably be more confident than the offspring of low-ranking ones.

In wild troops, the role of the peripheral males is to warn the alpha male and then to help him defend the troop. These males also play with, punish, and are groomed by juvenile males who, as Eaton remarks, "undoubtedly learn a great deal about macaque social behavior through this association." Peripheral males are sometimes allowed to join other troops and thus provide a genetic exchange which circumvents inbreeding. Since the Oregon troop is enclosed and not threatened by strangers, the strict social barriers which confine the peripheral males to the fringes of the feral troop are less stringent; as a result, they are generally accepted and are harassed only occasionally.

Several females in the Oregon troop have adult sons, eight to ten years old, who are still defended by their mothers;

why these adult males still depend on their mothers' defense is not clear. Brothers and sisters generally defend one another and their mother. After puberty, however, the males no longer defend their mothers as assiduously as before, but a strong bond between mother and daughter remains throughout life. Mothers and daughters spend much time grooming each other.

The adult females of the Oregon troop form alliances with one another, but unrelated males rarely do. This may explain why some feral males are driven away from the troop and why some are low ranking. When a male fights with a female, the other females are likely to come to her assistance, but adult males rarely assist one another. Such strong female alliances are believed to contribute to the fusion of feral troops. Japanese observers reported that continual fighting among groups of females within a troop precedes the separation of the troop. New troops generally consist of entire families which have broken away from the old troop. Thus, some have suggested that female alliances are more important than sexual attraction to the cohesive social order of Japanese macaques since such attraction in this species is mostly seasonal. Some males form an alliance with females and presumably defend them. One female in the Oregon troop, who was "adopted" by an old subleader male when her mother died, is still being cared for and defended by him even years after adoption. This male never mates with her probably because of the near father-daughter role. Another male friend, however, does mate with her and forms a loose consort relationship during the fall and winter mating season; the older male defends her during the nonbreeding season.

The mating season for these Oregon macaques occurs during October and November and most of the young are born during April and May. The summer is a relatively quiet season when the females are preoccupied with nursing their infants

and grooming each other and the males and the latter are engaged in play. With the approach of autumn, however, the males begin to swagger (Figure 35), holding their stubby little tails erect, and to indulge in several courtship "displays" such as shaking objects, leaping up and down, and drumming their feet on the ground. The males that display most vigorously also mate most frequently. As time goes on and the mating season approaches, the males become aggressive (Figure 30), threatening and attacking the females and fighting among themselves. They also begin to groom the females more frequently than during the nonmating season. Gradually some loose consort pairs are formed which stay together for a few hours, seldom longer than a few days. When the pair breaks up, each becomes a partner in a new alliance. Most of them mate with several different partners during the breeding season.

Social rank is not correlated with mating. Although the alpha male is no mean performer, one of the lowest ranking males mates much more frequently than he does. In the mating game, rank counts for nothing, and prestigious females seldom rebuff low-ranking males. In a free-ranging society, the males initiate courting behavior but the females choose the partners. Several females in the Oregon troop have even refused Alpha I, passively, to be sure, by not presenting when he indicated he was ready for action. When so rebuffed, Alpha I usually follows the female for several days, apparently unwilling to admit defeat. Eventually, however, he gives up and showers attention on a willing partner. In the meantime, the recalcitrant female generally mates with several other males whom she finds more acceptable. Similar situations with dogs and cats have been reported. Since paternity cannot be established easily, it is impossible to know which males contribute most to the gene pool. Apparently the dom-

inant males are so preoccupied with their civic responsibilities that they have a proportionately lower number of matings.

The increase in male aggression during mating time is probably triggered by a rise in male androgen levels. But the relation (if any) between testosterone and aggression is conditioned by social factors. No correlation between a male's social rank and his testosterone levels has yet been documented. Thus, although testosterone appears to elicit increased aggression in males during the mating season, it is not as important as social stimuli in doing so. In other words, male dominance and hierarchy are less dependent on size and testosterone levels than on other complex, less obvious factors. The alpha male of this troop, for example, is a smallish adult with no canine teeth and only one eye; yet his authority has so far never been seriously challenged. If he attacks another male, the latter does not fight back but only attempts to escape. In the Oregon troop, physical equipment, such as brute strength and canines, plays little or no role in attaining or maintaining high rank.

Consort pairing terminates with the end of the mating season. Males who fight viciously during the mating season are more likely to wrestle and chase each other playfully and even to play with juvenile males. Young males take advantage of the liberty granted them by adult males and indulge endlessly in play-biting. During this time, too, the once vicious adult males express a high degree of parental concern for the juveniles, grooming, defending, huddling with, and carrying them on their backs. This paternal concern is unrelated to dominance rank in our troop, but in the Takasakiyama troop in Japan, Junichiro Itani of Kyoto University believes it is displayed only by high-ranking males. This is not surprising since even in free-ranging troops in Japan, different troops have different cultural patterns. The troop in Koshima Island

washes sand from their food, those elsewhere do not. Even sexual behavior may differ from troop to troop. For example, in some troops female macaques court the males with symbolic mounts. In other troops, only those with a certain rank do so; and in still others, none do.

Beginning in the 1970-71 winter, the Oregon troop began rolling large snowballs on the ground. These objects later become the center of attention for infants and juveniles who played on them and for adults who sat on them. The snowballs had, of course, no functional significance to the macaques and represent only another instance of the uniqueness of each troop's level of cultural patterns, the product of their own special intelligence and skills.

Occasionally, for no apparent reason, members of the Oregon troop single out one animal and proceed to mob it. This action does not appear to be caused by crowding in the corral because although experiments with crowding the animals into a small enclosure resulted in increased mobbing, this apparently was due to removal from a familiar habitat rather than to increased population density. Crowding per se produced only a mild increase in aggression during which no breakdown in social structure occurred and the importance of the dominance hierarchy became more obvious.

Despite a population explosion (from the original 46 animals in 1964 to 200 members in 1975), the Oregon troop of Japanese macaques thus far gives no indication of abnormal behavior. On the contrary, a significant decrease in aggression was observed among the adult males between the 1971-72 and the 1972-73 mating season when the troop had increased from 107 to 125 individuals (a 17 percent increase in population density). The decline in aggression during the 1971-72 season, however, was associated with a change in the hierarchy which involved the six highest ranking males; by 1972-

73, when the new hierarchy had stabilized, there was more aggression. From these observations, Eaton concludes that the Oregon monkeys are like people in their response to increasing population density because in human societies, a stable social structure is a critical factor in defusing aggressive behavior.

A second two-acre corral has just been built; it connects by a tunnel to the first, and this gives rise to speculation about the effect on the Oregon troop. We anticipate that the troop will split in two. But, if fission occurs, will it happen when the troop's social structure is collapsing? Which maternal lineages will emigrate? Who will be the alpha male? Which males will be subleaders? What effect, if any, will the old dominance relationships have upon the new social order? What will happen to the old social order if and when key members leave? These animals will offer behavioral scientists a unique opportunity to study the genesis of a new social order.

Since members of the genus *Macaca* are, next to man, the most widely distributed existing primates, they must be considered to be the most successful. An understanding of their complex social order, therefore, and of their manner of communication may provide an insight into their adaptive success and may ultimately contribute to an understanding of man's adaptation to different environments and to his ultimate survival.

CHAPTER 5

DISEASES COMMON
TO MONKEYS AND MAN

Having seen that man shares his most important biological properties with nonhuman primates, we can better appreciate why man should share or be susceptible to the same diseases as his poor relatives and why some monkeys at least are good substitutes for man in the study of specific physiological parameters as well as of certain diseases. Such exceptions as leprosy do not invalidate the rule. Often, however, there are important differences in the clinical progression of the same disease in man and monkeys.

The B virus, for example, which is usually innocuous to rhesus monkeys, is virulent in and even fatal to man. On the other hand, the *Herpes simplex* virus, which generally causes only fever blisters in man, is fatal to rhesus and other species of monkeys. Squirrel monkeys are in the pink of health even when their blood crawls with still undefined plasmodia, filaria, and trypanosomes. Many common diseases are easily transmitted by the same vector from monkey to man and vice ver-

NOTE: This chapter is based on a lecture in the Wesley W. Spink Lectures on Comparative Medicine presented at the University of Minnesota, Minneapolis campus, on October 17, 1975.

sa. A few years ago an epidemic caused by an undescribed high-morbidity, low-mortality pox virus swept through our colonies of rhesus and Japanese macaques (McNulty et al., 1968); several handlers were infected with it and showed ulceration pocks identical with those in the monkeys (Figure 36).

I shall limit my brief discussion to only a few common diseases. The literature about each of these disorders is so voluminous that a comprehensive discussion or review would seem to be redundant. I have chosen my examples principally to elucidate the exceptional contribution made by nonhuman primates to man's knowledge of the epidemiology of some diseases that have plagued him for centuries.

Tumors

Despite evidence to the contrary, scientists still believe that man is much more susceptible than monkeys to cancer. The literature on tumors in nonhuman primates contains numerous references to assorted species of primates of unknown origin and age, lumped together under the vague term *monkeys*, a common, deplorable practice in much of the scientific literature. Worse still, the findings have been compared with the incidence of tumors in man without regard to biological relationships, which are more remote between some "monkeys" than between man and those primates that most resemble him.

During the past two decades, systematic observations and more scrupulous search have clearly shown that neoplasms are not rare in nonhuman primates. These findings raise a number of relevant questions: (1) Do the morphology, location, natural history, frequency, and age incidence of spontaneous malignant neoplasms in nonhuman primates resemble

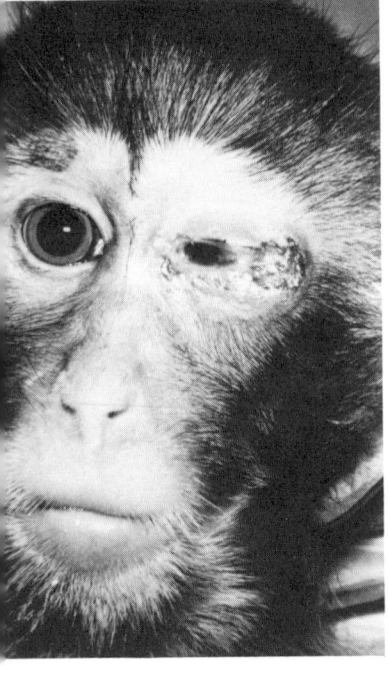 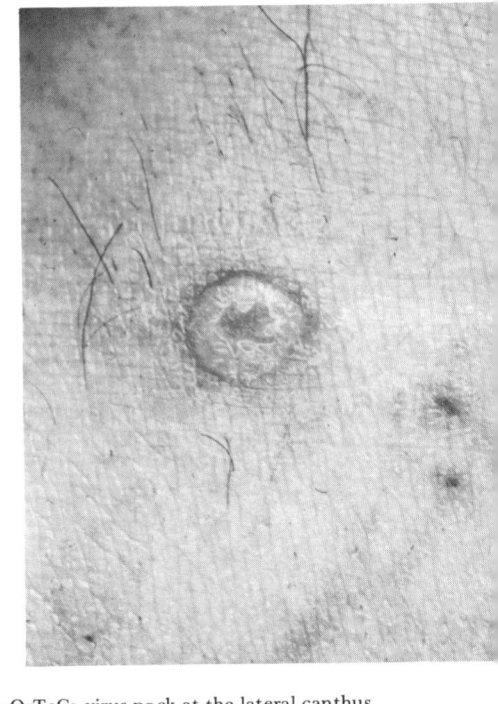

Fig. 36. a. *Macaca mulatta.* Spontaneous OrTeCa virus pock at the lateral canthus of the eye. (Courtesy of Dr. W. P. McNulty. Figure 15a from McNulty, 1972, in *Pathology of Simian Primates*, Part II, R. N. T-W-Fiennes, ed., reprinted with permission of S. Karger AG, Basel; Figure 16 from McNulty et al., *Archives of Dermatology* 97:286-293, copyright 1968, American Medical Association.) b. Spontaneous OrTeCa pock on the flexor surface of the forearm of an animal handler. (Courtesy of Dr. W. P. McNulty. Figure 16 from McNulty, 1972, in *Pathology of Simian Primates*, Part II, R. N. T-W-Fiennes, ed., reprinted with permission of S. Karger AG, Basel; Figure 1 from McNulty et al., *Archives of Dermatology* 97:286-293, copyright 1968, American Medical Association.)

those in man? (2) If there are similarities, are they correlated with the taxonomic affinities of the animals? (3) How do the different species of primates respond to carcinogens compared with both man and the common laboratory animals most widely used for such studies?

Because malignancy is overwhelmingly a disease of old age and because most nonhuman primates have long life-spans in captivity (macaques, the most widely used experimental primates, can live thirty or more years), cancer studies are long

and tedious. The answers to our questions, therefore, though important, are still tentative. The pressures of other research interests on a limited resource of animals and the inordinate expense of maintaining monkeys for decades make it impossible to keep large numbers of primates in the laboratory until they reach old age. Furthermore, the many species of primates and our partially successful methods of animal husbandry have made it necessary to restrict our studies on carcinogenesis to only a few species.

For this reason, the findings at the Oregon Regional Primate Research Center, as well as in other laboratories, have been made chiefly on rhesus monkeys and, to a lesser extent, on baboons, Celebes black apes, Japanese macaques, and such prosimians as fat-tailed galagos (*Galago crassicaudatus*), ringtailed lemurs (*Lemur catta*), and fulvus lemurs (*Lemur fulvus*).

Of the approximately 900 rhesus monkeys now alive at the Oregon center, only 270 were born in captivity and are, therefore, of known age. Of these, only 59 are older than ten years, and none are over twelve; that is, they are all in the first third of their life-span. And yet, in the total accession of 4,800 macaques in our population of twelve years, eight malignant neoplasms have been found: two renal carcinomas, two lymphomas, one seminoma, two soft-tissue sarcomas, and one choriocarcinoma. Not only is this incidence consistent with that found in an equal population of young and early adult human beings, but the tumors are histologically similar to those in human patients. The singular absence of leukemia from our colony remains unexplained. Furthermore, other laboratories have reported isolated cases of the most common human malignancies—carcinomas of the gastrointestinal tract, breast, skin, uterus, and prostate—none of which have been found in our colony. The lack of these tumors in our animals

is, of course, understandable since these types of cancer characteristically occur in middle- and old-aged human beings.

If the prosimians are grouped together in a single category dominated numerically by fat-tailed galagos (*G. crassicaudatus*), fulvus lemurs, and ringtailed lemurs, ten malignancies have been observed in about 500 animals: one each in the pancreas, liver, and kidney, two each in the lungs and uterus, and one fibromyxosarcoma, one malignant mesenchymoma in the axilla, and one leiomyosarcoma in the chest. The histologic appearance of these tumors, however, unlike those in macaques, is different from that of comparable human tumors. Any attempt to make meaningful speculations about them and about the age incidences in prosimians would be futile because we do not know their exact life-spans and our small population contains only a few wild-caught, apparently old individuals.

However meager, these few data are at least consistent with the theory that malignant tumors occur throughout the Order Primates with the same types and patterns of incidence as in man.

Radiations and chemicals known to be carcinogenic in laboratory rodents, and perhaps even in man, also cause neoplasms in both simian and prosimian primates. At the Oregon center, chronic painting of the back skin of rhesus monkeys and fat-tailed galagos with dimethylbenzanthracene and dodecyl benzene has produced multiple papillomas, basal cell tumors, and several dermal sarcomas four to six years later. Thus, as in man, the tumor induction period was much longer than in laboratory rodents. Apparently, the "clock" which sets the rate of the aging processes also governs the speed of carcinogenesis. There is, obviously, an urgent need to continue such studies in a systematic way; the investment of time and money will pay dividends in long-range results.

Arterial Disorders

So many middle-aged as well as old Americans die annually of cardiovascular disorders that survivors by the thousands, frightened by the statistics, have become fanatic joggers and cyclists. Considering the millions of people, young and old, throughout the world who are living with diseased arteries which went undetected in their early stages, one can be at least sympathetic with the hordes who have taken to the cinder and bike paths. But despite their exertions, many of these people are doomed to develop ischemic hearts which may eventually prove fatal. What is even worse, until precise methods for detecting the early progress of coronary atherosclerosis are discovered, men will continue to be the victims of the advanced form of this subclinical disease. In other words, we need a cure, or better, a preventive, for the disease and we need it *now*. Thus, we can appreciate the urgency of finding a suitable experimental animal in which to induce a type of coronary atherosclerosis which is similar to that in man. In our search, we must focus on those animals in which we can follow the induction, progression, and regression or arrest of atherosclerosis. The animals of choice are, of course, nonhuman primates because in all species studied thus far various types of atherosclerotic lesions have been reported. (The fact that some investigators still prefer to use chickens, Japanese quail, rats, rabbits, and other species is one of the more obscure mysteries of modern research.) But before experimental studies can begin on these primates, we need to know as much as possible about the occurrence and types of the disorder in wild, free animals, lest the wrong species of animal be chosen. In all of the surveys made thus far, both in Old and in New World monkeys, so little has been learned about the incidence and prevalence of the spontaneous form of atherosclerosis that even generalizations are unwarranted.

The spontaneous lesions that have been found are generally so small that they do not seem to impair normal arterial functions as they do in man.

Like the lesions found in the arteries of wild baboons (McGill et al., 1960), those in feral howler monkeys have been found to resemble the early stages of atherosclerosis in man (Malinow and Maruffo, 1966); Malinow and Storvick, 1968). Unlike human atheromata, however, which in Caucasians and American blacks occur predominately in males, those in nonhuman primates are not related to sex. Moreover, they occur in animals whose diet contains little or no cholesterol and which have relatively low plasma cholesterol levels. Nonetheless, Gresham (1973), who believes that these lesions are ideal for studying the evolution of atherosclerosis, points out that similar lesions in children may be the substrate upon which, as a result of diet and rising blood pressure in later life, lipids collect and eventually occlude the arteries.

The major shortcoming of all studies of atherosclerosis in experimental primates is that the condition proceeds at a much more rapid rate in them than in man (McNulty and Malinow, 1972). As a result, the conditions induced over a brief period of time in young, healthy animals cannot be compared with those found in older human patients, which have developed over a period of years. Moreover, the scientist who performs experiments on laboratory primates must consider that most of these animals were only recently trapped, incarcerated in an impoverished, hostile environment totally different from their natural one, and given diets which though perhaps nutritionally superior to their food in the wild are foreign and often eaten only because nothing else is available. All animal attendants know that caged animals, living as they do in a constant state of stress, are frightened and tend to develop neurotic behavior. Even under the very best laboratory supervision, these and other conditions cannot be avoided,

but they can and often do raise cholesterol levels. Before these increases can be exactly assessed, it is imperative that the incidence and quality of the naturally occurring vascular disorders in feral animals be known.

Perhaps because few nonhuman primates are allowed to reach old age, they rarely show the obvious clinical signs of cardiovascular disease (McNulty and Malinow, 1972). The actual age of most adult captive and feral animals can only be guessed at, and the potential life-span of most species is an even darker mystery. Comprehensive autopsy records of the arteries of feral animals are available only for rhesus, cynomolgus (*Macaca fascicularis*), and howler monkeys (*Alouatta caraya*) and for baboons; a few scattered records have been kept for chimpanzees and squirrel monkeys. But even such scant data indicate that we know a great deal more about the incidence and type of atherosclerosis in man than in these nonhuman primates. Moreover, since the diagnosis of minimal disease depends on the size of tissue samples, diagnostic criteria, and the acumen of the observer, reports about the incidence of spontaneous arteriosclerosis in monkeys vary widely. There is too little evidence to determine whether these abnormalities have much bearing on the health or longevity of the animals (McNulty and Malinow, 1972). Most of the reports on the incidence of cardiovascular pathology in primates are based on zoo records and are, therefore, sketchy. Since the heterogeneous groups of nonhuman primates cannot be categorized simply, the incidence of cardiovascular disease in the various species cannot be compared. Nonetheless, the numerous subclinical alterations found in the heart and vessels of nonhuman primates may be helpful in elucidating the causes and treatment of the disease in man.

In recent years, much of the work has focused on the experimental induction of atherosclerosis in laboratory animals, and the resulting pathologies have been discussed and com-

pared with the naturally occurring diseases in the same species. The development of arterial lesions has been followed in animals on diets with high levels of cholesterol, but not enough is known about how this and other substances are metabolized by many experimental animals. Lipids have been studied assiduously, more so than any other substance, and the results indicate that several primates metabolize them almost like man. More recently, nonhuman primates have been used to investigate some subtle biochemical mechanisms associated with the development of atherosclerosis. These include the ways in which composition and metabolism of arteries change with age and atherosclerosis and the metabolism of certain lipoproteins which are important in atherogenesis. In these respects, nonhuman primates resemble man much more than do nonprimates (Portman, 1970; Portman and Illingworth, 1975). But the stress of captivity alone can raise blood lipid levels in nonhuman primates and can, no doubt, alter many other biological parameters relevant to the development of atherosclerosis. Moreover, the stress phenomenon can vary from animal to animal and the investigator cannot afford to ignore divergences when he calculates his results.

The incidence of cardiovascular diseases will probably increase in man as his life-style becomes more stressful. This fact forces us to heed the urgency of knowing at least the etiology and progression of these diseases. The availability of nonhuman primates may make the acquisition of such knowledge more possible.

Cholesterol Gallstones

At the Oregon primate center, Osuga, Portman, and their colleagues (1971, 1972) found cholesterol gallstones in squirrel monkeys that were used for experiments on the metabolism

of lipids and lipoproteins in the arterial walls and in blood. The relation between gallstones and atherosclerosis in man is not clear, but two circumstances appear to relate them: the ectopic deposition of cholesterol in both conditions and the unusual prevalence of both in prosperous societies. Nonetheless, the two conditions are sufficiently different to cause one to question whether they are related. For example, the incidence of gallstones in women is almost twice that in men whereas men have a higher incidence of myocardial infarction, a manifestation of atherosclerosis. Another unexplained fact is the high incidence of gallstones in young American Indian women, who have relatively low plasma cholesterol levels. Moreover, some widely used therapeutic approaches to gallstones and atherosclerosis can exert different and even opposite effects. Contraceptive steroids and estrogens increase the risk of clinical gallstones more than twofold in young and postmenopausal women respectively but apparently have no effect on atherosclerosis. Unsaturated fats and drugs used therapeutically to lower cholesterol levels as a protection against atherosclerosis may increase the incidence of gallstones in man as well as in squirrel monkeys. Yet, chenodeoxycholic acid and other potentially therapeutic agents against gallstones could possibly aggravate the course of atherosclerosis.

Portman et al. (1975) found that the incidence and mass of gallstones in squirrel monkeys were greatly affected by diet. None of the monkeys on a standard diet had gallstones, but monkeys on a semipurified diet with 45 percent of the calories derived from butter and some added cholesterol developed gallstones. This last level of fat, incidentally, is similar to that normally ingested by people in the more affluent parts of the world. Obviously, dietary cholesterol favored the formation of cholesterol gallstones in monkeys. To identify

the factors that affect gallstone formation, Portman varied the type and level of fat, included or removed cholesterol, varied the source of carbohydrate, and added cellulose fiber to the standard diet. Of all the diets tested, those which contained cholesterol caused the formation of gallstones in a large number of animals. Even the standard semipurified diets with low levels of fat and no cholesterol were somewhat lithogenic. When some highly unsaturated fats were added to the cholesterol diet, the incidence of gallstones was at least as high as when butter was added to the same diet, but diets with high levels of saturated or unsaturated fat without cholesterol formed no gallstones.

The susceptibility of individual animals to these different regimens varied widely. Only half of the animals on one lithogenic diet in a series of studies developed stones. Furthermore, when the gallbladders of all of the animals on a lithogenic diet were drained, only the animals which had stones later formed new ones.

The value of these and other studies on gallstones and on their relation to atherosclerosis in squirrel monkeys is based on what happens in man. Gallstones in the two species are characterized by both similarities and differences (Osuga et al., 1974a, b): in the morphology of the stones and of their smaller constituent parts in the bile, in the composition of bile in subjects with and without stones, and in certain aspects in the formation of bile.

In monkeys, the number and weight of the stones were proportional to the length of time the animals had been on a lithogenic diet. As the large stones grew, their individual shapes became increasingly complex and new ones appeared to form. An examination of the stones and their precursor particles in the bile helped to determine the sequence of their development. In fractured surfaces that go through the center

of simple round stones, crystalline structures appear to radi-
ate from the center (Figure 37). Chemically, the gallstones of
these monkeys are largely composed of cholesterol, and the
shape of the plates they contain suggests at once that this is
free cholesterol. Gallstone formation proceeds from the
growth of very small crystals to successively larger ones and
then to an aggregation of microliths which form a skeleton
for the radial growth of the macroscopic gallstone. The simple stones of squirrel monkeys are microscopically somewhat
different from human stones which tend to develop a solid
amorphous center.

Stones in man and monkeys differ in the way in which the
larger complex gallstones are formed. In squirrel monkeys,
large multilobar stones develop by the aggregation of several
simple stones which adhere to one another partly by means
of the crystalline projections from their surfaces. In both
squirrel monkeys and man, complex gallstones resemble mul-
berries, with several simple and complex stones embedded in
a solid amorphous matrix. Cholesterol gallstones in man usually abound in bile pigments and various organic and inorgan-
ic salts, including those of calcium, but in the stones of squirrel monkeys, Osuga and Portman have so far identified only
calcium bilirubinate.

For a long time, investigators have suspected that cholesterol gallstones are formed in man when the concentrations
of cholesterol become too high in bile and that in some way
other constituents of the bile prevent cholesterol crystals
from forming. But only in the last few years has a physical-
chemical hypothesis been formulated to explain how cholesterol is crystallized from bile (Small and Rapo, 1970). Thus,
the presence or absence of gallstones in man can be predicted
on the basis of whether or not the bile is saturated with cholesterol. But the biliary composition of monkeys varies even

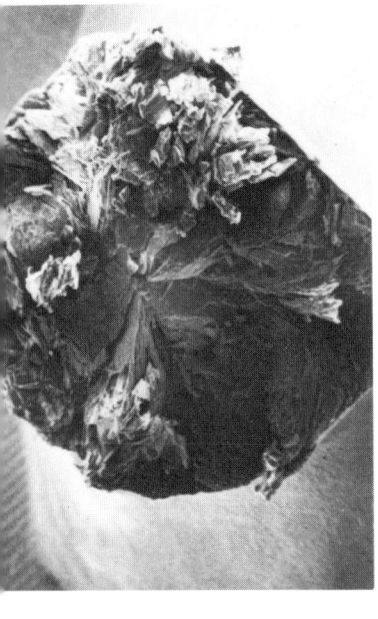

Fig. 37. a. Fractured surface through the approximate center of a medium-sized radial stone formed after six months on the lithogenic diet. Note the radial crystalline arrangement. b. Enlarged view. (Figures 13 and 14, respectively, from Osuga et al., 1974, *Laboratory Investigation* 30:486-493; U.S.-Canadian Division of the International Academy of Pathology, reprinted with permission of Williams and Wilkins Company.)

within a single dietary group with a high incidence of gallstones; that is, the bile of animals with stones tends to have higher cholesterol levels than that of animals without stones. Thus, genetic differences in bile composition may help to explain why certain animals consistently form stones and some are consistently free of them.

A note of caution should be sounded here. Despite the fact that squirrel monkeys and men are similar in bile composition and biliary physiology, neither the cause nor the prevention of gallstones in monkeys may be applicable to man. Dietary cholesterol plays a much more important role as a determinant of plasma cholesterol levels in squirrel monkeys than in man. This casts some doubt upon the importance of

dietary cholesterol in the formation of human stones. In both man and squirrel monkeys, the saturation of dietary fat greatly influences the levels of plasma cholesterol. Highly unsaturated safflower oil lowers the blood cholesterol in squirrel monkeys just as it does in man but when taken with dietary cholesterol results in a high rate of gallstone formation. In one study on human beings, the incidence of gallstones increased when safflower oil was used as the major dietary fat (Sturdevant et al., 1973). When the drug chenodeoxycholic acid, a widely tested agent that may dissolve gallstones in man, was administered to monkeys on cholesterol and either a saturated or an unsaturated fat diet, a high incidence of gallstones was observed.

Thus, whereas there are similarities and differences between cholesterol gallstones in man and squirrel monkeys, the differences in stone morphology, bile composition and physiology, and response to diet may not be any greater than those found among the individuals in any single species. The differences in biliary physiology and susceptibility to cholesterol gallstones within the Order Primates — from rhesus monkeys, which have resisted all experimental efforts to induce gallstones, to some baboons which form them spontaneously (McSherry et al., 1971) — are the most perplexing aspects of the problem. Before adequate animal models of human disease can be developed, such individual and species differences must first be explained. Perhaps the clues to the disease lie at that basic level of investigation.

Diabetes Mellitus

About 2 to 3 percent of the affluent human population worldwide has developed or will develop diabetes, a disease best characterized as one of impaired capacity to metabolize

glucose coupled with rapidly progressing vascular disease. Clinical signs include polyuria, high susceptibility to infections, impaired renal function, vascular disorders, weakness, and sometimes coma. Laboratory findings include glucose intolerance, insufficient insulin along with normal or increased glucagon, and increased levels of plasma triglycerides and pre-betalipoproteins. The pathologic findings include atherosclerosis, microaneurysms of muscle capillaries and retinopathies, loss of weight, opacity in the ocular lens, and often partial or total destruction of beta cells in the pancreatic islets of Langerhans. These and other signs typical of human diabetes have been observed to occur spontaneously in numerous species of experimental mammals, notably in selected strains of mice, hamsters, and rats. The record of its occurrence in nonhuman primates has been fragmentary, and the details of its course have been minimal. What data we have could indicate as much as a 1 percent incidence, but since few laboratories have compiled any data on large numbers of animals, this figure should be taken *cum grano salis*. Furthermore, the subtle early signs of diabetes are difficult to detect in animals and can easily escape notice unless the animals are under constant and expert surveillance. Even in severely diabetic animals, such obvious signs as lethargy, polyphagia, polydipsia, and polyuria can be missed except by the trained observer.

Sketchy reports of the detection of diabetes in nonhuman primates date back to the last century but they are mostly anecdotal. Apparently the first documented case was of a postpartum female Hamadryas baboon in 1945. This animal showed clear-cut clinical signs of the disease, including glucosuria, hyperglycemia, and an abnormal glucose tolerance test (cf. Howard, 1975). Since rhesus monkeys have been used as experimental animals more widely than other primates, we have more comprehensive information on their health prob-

lems than on those of other primate species. Several spontaneously diabetic cases have been reported among rhesus monkeys, particularly among postpartum females. When rhesus
mothers give birth to abnormally large infants (greater than
700 g), this serves to alert personnel monitoring a large colony of breeding primates to the possibility of a diabetic mother. Diabetes mellitus has also been clinically diagnosed among
prosimians, in various species of macaque, and among the
great apes, in chimpanzees and orangutans.

In addition to the primary signs of the disease, man develops numerous secondary complications — e.g., atherosclerosis, microangiopathies, glomerulosclerosis, and cataracts.
These, however, occur much less frequently and with much
less intensity in nonhuman primates.

Lacking observations from the field, we do not know
whether the diabetes detected in laboratory animals is actually spontaneous or is induced by such stresses as confinement,
handling, and changes in diet and environment. However,
since some signs of the disease have been detected shortly after their capture, these animals may have a natural predisposition to diabetes. Howard (1974a, b) suggests that spontaneous diabetes in nonhuman primates may reflect a "genetic
weakness" as it does in man.

While engaged in a project to induce diabetes in monkeys,
Howard (1972a, b) became aware of a species in which some
monkeys possibly had spontaneous diabetes. Examination of
the clinical records of the small colony of Celebes black apes
(*Macaca nigra*) (Figure 38) at the Oregon Regional Primate
Research Center showed many with apparent hyperglycemia.
Since blood glucose of man and most other primates ranges
from 60 to 100 mg/dl, consistent values of 130 mg/dl and
higher indicate the abnormality of hyperglycemia. A number
of the Celebes black apes had inappropriate hyperglycemia,

Fig. 38. A diabetic Celebes black ape (*Macaca nigra*).

and the diagnosis of diabetes was corroborated by their inability to clear glucose in a tolerance test performed intravenously. These signs, coupled with increased blood lipids and decreased circulating insulin, led to the diagnosis of moderate to severe diabetes, or more conservatively, these monkeys are said to have a diabeticlike syndrome. Since to our knowledge no other systematic studies of an entire breeding colony have been made before, Howard's discovery of a diabetic-prone colony of Celebes black apes marks an important advance: it enables him to pursue not only the genetic aspect of the disease but its development and progression from birth as well.

Howard's colony of fifty animals of various ages has the highest known incidence of diabetes ever reported in nonhuman primates. It has been diagnosed in four- and five-year-old juveniles of both sexes just reaching sexual maturity as well as in adults; over 50 percent of the colony is involved. That this appears to be species-specific has been corroborated by the appearance of about the same percentage of abnormalities in a smaller colony at the Yerkes Regional Primate Research Center, among animals at the Seattle Zoo, and in isolated cases among Celebes black apes in several other zoos. The severity and frequency of signs in all of the animals were similar. Some of the older or more severely diabetic animals have died and have been replaced.

Histologic observations of the islets of Langerhans disclosed changes often seen in human juvenile diabetics. The pancreatic islets in all Celebes black apes with ranging severities of diabetes were variably infiltrated with amyloid, which destroys the insulin-secreting beta cells. Similar amyloid infiltrations observed in the pancreas of other diabetic nonhuman primates are more extensive than in human diabetics. The disease in these animals has all the characteristic signs of human diabetes. As a monkey becomes increasingly diabetic, insulin

therapy is often necessary. Control with insulin injections seeks to restore blood sugar concentration to normal in man, but control in nonhuman primates is much more difficult. Since complete amelioration of hyperglycemia and glucosuria with insulin in these monkeys is so difficult that fatal overdoses can easily be administered, investigators compromise by administering insulin sparingly. Although this fails to bring the disease completely under control, it does make possible a greater manifestation of the secondary complications.

Why diabetes is so species-specific as to attack a high percentage of *Macaca nigra* that have only recently been feral is difficult to explain. Howard (1974a, b) suggests that these monkeys, which inhabit a small sector of the Celebes Island in the Southwest Pacific, "have inbred over hundreds of thousands of years and eventually developed a genetic weakness or predisposition towards the diabetic condition. Thus, when they are removed from their natural environment, caged, and fed a commercial diet with high concentrations of carbohydrate, those with the genetic weakness tended to develop the diabetic signs." As he admits, however, without field data such a hypothesis can only be conjectural.

This is an excellent example of a common human disease which also occurs in nonhuman primates. Compared with the vast amount of knowledge we have on human diabetes, however, the data on other primates are remarkably scarce. This fact alone constitutes a mandate to extend and deepen our study of this disorder since only when we have acquired a comprehensive view of its biological properties in laboratory animals, including nonhuman primates, will we learn how to control, perhaps even to eradicate, it in man. At the moment, the Celebes black apes seem to offer some hope that such a goal is attainable.

Yellow Fever

Formerly one of the world's worst diseases, yellow fever is endemic to Africa and the tropics of the New World, where it was well known even in the seventeenth century. The discovery of its viral etiology and of a mosquito vector led to the postulation of the so-called "Gorgas doctrine," that is, that since yellow fever cannot exist without mosquitoes, control of the insects would lead to its eventual eradication. A single virus is responsible for the two types of yellow fever, but the host and insect vectors differ for each. City dwellers who suffer from yellow fever are infected by the common mosquito, *Aedes aegypti*. In the jungle, where monkeys are the normal hosts, the disease is transmitted by a number of arboreal mosquitoes.

When Walter C. Reed was with the U.S. Army Yellow Fever Commission in Cuba in 1902, he injected filtered serum from a yellow fever patient into a normal human volunteer, who promptly came down with the disease. Thus, he showed for the first time that a filtrable agent, probably a virus, caused the disease and that the virus was usually transmitted by the mosquito *Aedes aegypti*. After the insect vectors were controlled, the disease was almost totally eradicated in most of the New World, and the construction of the Panama Canal became a reality. Incidentally, the control of mosquitoes also alleviated the incidence of malaria.

The knowledge that yellow fever is caused by a virus that grows in man and is transmitted by mosquitoes was only a first step. It then became imperative to find experimental animals which could also host the virus. During the ensuing search for the animal model, the West African Yellow Fever Commission discovered that after being injected with blood from a yellow fever patient, a rhesus monkey became in-

fected. Thus a second milestone in the search of ways to control yellow fever was reached. Quite by coincidence, the strain of the virus isolated from that rhesus monkey was by far the most widely used in research and eventually led to the development of yellow fever vaccine. The malady was found to be transmitted from monkey to monkey by the same *Aedes aegypti* that transmits the disease to man.

Outbreaks still plague the eastern shores of the Panama Canal from time to time. Since no enzootic diseases (i.e., diseases peculiar to animals in a given locality) could be demonstrated in eastern Panama, the periodic waves of yellow fever are thought to be epizootic extensions emanating from somewhere in South America. Despite the lack of agreement on this point, the fact remains that a yellow fever virus lurks somewhere east and south of the Panama Canal, sometimes affecting local monkeys and giving rise to widespread epidemics.

Most of the South American primates — marmosets and tamarins, squirrel monkeys, night monkeys, howler monkeys, spider monkeys, and capuchins — are susceptible to yellow fever, but their susceptibility and tolerance vary from species to species, from one individual to another, and with the strain of the attacking virus. According to Bugher (1951), all New World monkeys can be infected with the virus and can concentrate enough of it in their blood to infect mosquito vectors. Thus, monkeys are the major links in the chain that transmits yellow fever in the American tropics.

Laboratory research, in which monkeys are infected with a yellow fever virus, has yielded some pertinent data. The single genus of howler monkeys, *Alouatta*, which range south from Mexico to Argentina and Bolivia, are probably the most susceptible animal hosts. But because they are extremely difficult subjects to keep well or even alive in the laboratory,

relatively little experimentation on the progression of yellow fever has been done on them. Davis (1931), who infected one howler with mosquito bites and then transfused it with the blood of a rhesus monkey, recovered the virus in the rhesus and in mosquitoes that had been allowed to feed on the rhesus. Since then, several investigators have successfully carried cyclic transmissions from howlers to different species of *Aedes* mosquitoes and to other monkeys. Surveys have repeatedly shown that howler monkeys from Mexico, Central America, Panama, Trinidad, and South America are infected and are, no doubt, the natural repositories of the infecting organisms. All experimental work with the six genera of marmosets and tamarins from Panama to southern Brazil has shown that these species are highly susceptible to the disease which in them is accompanied by high morbidity and mortality. In South America, these animals may be important links in the transmission of the virus, but in Panama, the marmoset, which lives mainly in second-growth forests that are not ideally suited for sustaining high population densities of mosquitoes, seldom comes in contact with the vectors and may not be important in transmitting yellow fever.

The single species of squirrel monkey, which ranges from Venezuela, the Guianas, the Amazon area of Brazil, Colombia, the Pacific coast of Western Panama, and Costa Rica, is highly susceptible to the yellow fever virus and may have been infected through a long chain of monkey-mosquito-monkey passages. When infected, these little animals become very ill and usually die. Since they have a fairly restricted range, they are only locally important in the natural transmission and spread of yellow fever.

The single species of owl monkey which ranges from eastern Panama to the Orinoco and Amazon basins, is also highly

susceptible to infection. The high concentrations of virus in the blood of these monkeys easily infect mosquito vectors. Once infected, these monkeys suffer high morbidity and fatality, but because of their restricted distribution and nocturnal habits, they do not infect many diurnal vectors. Thus, they are not important in maintaining the transmission cycles of yellow fever.

The several ill-defined species of the single genus of spider monkey, *Ateles*, are found from Mexico to Brazil. In Brazil and Panama, they are easily infected but seldom die from the virus which they pass on to mosquitoes.

Like the *Ateles*, the capuchins of Central and South America are susceptible enough to the yellow fever virus and are capable of infecting mosquito vectors, but the infections are asymptomatic and the animals seldom die.

Although several Old World monkeys are also susceptible to yellow fever virus, they have not been used in laboratory research nearly as extensively as the New World Platyrrhines mentioned here.

Thus, from the first discovery of the yellow fever virus to its first transmission from man to nonhuman primate, monkeys have played an indispensable role in its characterization and control. They were used to clarify the differences between the urban and jungle types of the disease, and studies of the disease in monkeys led to the production of the first vaccine. Because New World monkeys are the major reservoirs of yellow fever, they are now used by public health authorities in surveillance measures in the jungles of Central America.

Yellow fever is only one of the hundreds of known viruses that threaten the health of human and nonhuman primates alike. In any consideration of public health problems, the lat-

ter's role in the solution of the yellow fever problem remains a classical example of the value of nonhuman primates to man's safety and well-being.

Malaria

Even though for centuries malaria has been one of the world's most widespread diseases, and still is, the parasites that cause it in man were discovered by Laveran only in 1880. Five years later malarial parasites were also found in birds, and in 1898 the incomparable Koch found them in monkeys. It was Koch himself who in 1900 first attempted to transmit the malaria organisms *Plasmodium vivax* and *P. falciparum* from man to orangutans and gibbons. The negative results led Koch to assume that the apes were not susceptible, but he suggested that animals not so closely related to man might harbor the parasites that cause human malaria. As it turned out, he was right. Although chimpanzees and gibbons are susceptible to human malarias only if splenectomized, several New World monkeys can easily be infected with all of the human parasites. Over the years, experimental infections have been transmitted between man and other primates by inoculation with infected blood and with infected mosquito vectors.

In India, both rhesus monkeys and man are naturally susceptible to a malaria caused by *P. knowlesi*. Since this organism in man produces a shorter and milder form of the disease than the four well-known human malaria organisms, the pathogens from rhesus monkeys were used to treat patients with neurosyphilis, which is almost always fatal but which responds well to periods of high fever. However, the former practice of treating one disease by superimposing another has now been largely discontinued.

Because of the ease with which monkeys can be infected with malaria parasites, they have been widely used to test the efficacy of compounds with potentially therapeutic value. Using select experimental animals to study human malaria also has other advantages. When the parasites were first discovered in human red blood cells, the pathogens were believed to go through two distinct states: one causing the characteristic fever, the other, which is harmless, infecting mosquito vectors. According to this hypothesis, the parasite then goes through a third stage in the mosquito and when passed back to man through the mosquito's proboscis causes the disease. The lag period between the transfer of this third-stage parasite to man and its appearance in his red blood cells suggested that still another stage of parasite had to be developed somewhere in the body. This proved to be the case when the elusive fourth stage was observed developing in the liver cells of infected rhesus monkeys. Eventually all four species of human malaria pathogens were also found to go through the fourth or liver-cell stage in rhesus monkeys. We now know that after a first attack, this fourth stage can remain asymptomatic indefinitely until the pathogens are released into the blood stream and so-called relapses of the disease occur.

The progression of these studies is almost classical: first human malaria and then monkey malaria were studied in man, and finally, monkey malarias were studied in monkeys. The discovery of more and more adequate drugs for treating malaria demonstrated the effectiveness of these investigations. Better methods of controlling the spawning of the mosquito vectors were devised for malaria organisms.

All went well for a time until a new population of malaria-bearing mosquitoes resistant to DDT, malathion, and other insecticides replaced the susceptible population, on the one hand, and drug-resistant malaria parasites replaced the sensi-

tive parasites on the other. Thus, no sooner had the original problems been solved than new ones appeared and new investigations had to be launched. What was needed was an animal whose susceptibility to malaria organisms was predictable and in which more effective therapy could be developed. To be practical, such an animal should be easy to obtain, to keep healthy, and to breed in captivity. Even though such an animal does not exist, one has been found which comes close to it.

In 1966 Martin Young and his colleagues at Gorgas Memorial Laboratory in Panama successfully transmitted the common human *P. vivax* pathogen to the small night or owl monkey (*Aotus*). Later, Quinton Geiman injected a small amount of blood from a patient with *P. falciparum* into a splenectomized owl monkey (cf. Schmidt, 1973). Some three months later parasites began to appear in the blood of the recipient monkey, and as time went on the number of parasites increased steadily until Geiman transferred them from the original monkey to a second splenectomized one. This new animal developed an even more severe form of the disease and with repeated passages to other owl monkeys the infections became increasingly fatal, especially in animals with intact spleens. Later Geiman succeeded in infecting owl monkeys with several strains of *P. falciparum*. Notwithstanding these successes, the nocturnal owl monkey made a poor laboratory model. It was notoriously fragile and reproduced only with help. These problems were reviewed by the National Center for Primate Biology at Davis, under the direction of an epidemiologist, L. H. Schmidt, who has had a long-standing interest in both human and simian malarias. After Schmidt received his first shipment of fifty owl monkeys, all but two promptly died from an infection of *Herpes simplex*, which, as mentioned earlier, in man normally produces only a harmless

blister. Dr. Schmidt learned much from this disaster, including a way to avoid *Herpes* infection (personal communication). Having overcome with remarkable dispatch the first important hurdle, which was one of husbandry, Schmidt obtained from human subjects those strains of *P. falciparum* which had demonstrated widely different responses to antimalarial drugs. Some were susceptible to all drugs used, some were resistant to all drugs, and some were resistant only to specific drugs but susceptible to others. Using this wide spectrum of strains, Schmidt first determined the characteristics of the diseases in owl monkeys and then proceeded with the almost endless task of screening for effective antimalarial drugs. Several of these have been found, but Dr. Schmidt and his colleagues work on.

Maintaining owl monkeys in the laboratory still poses many problems. To begin with, the animals cannot be purchased easily from the none-too-friendly South American nations, which have placed partial or full embargoes on nearly all of their native nonhuman primates. Unless these embargoes are lifted, we may have to dispense with the services of these potentially splendid animal models in studies on malaria, a disease that may be with us for many years to come. Here is another instance of one of the gravest problems confronting us in biomedical research: the scarcity and high cost of nonhuman primate models.

Tuberculosis

Finally, we come to the disease to which monkeys are most devastatingly susceptible and against which all reputable nonhuman primate colonies must maintain constant surveillance. Whether tuberculosis occurs naturally in New and Old World primates is difficult to say, but most laboratory species, in-

cluding the great apes, are highly susceptible to both the human and the bovine as well as the avian bacilli. Rhesus monkeys are so sensitive to both that infections spread swiftly in colonies and the animals invariably die. Tuberculosis is principally pulmonary in this species, but lesions have often been found in the gastrointestinal tract, liver, spleen, and lymph nodes also. Not all species of primates studied, however, are equally prone to infections. Early surveys indicate, for example, that the mangabeys may be the most susceptible and the prosimians and marmosets the least susceptible with the rhesus monkeys intermediate (Kenard et al., 1939). Despite such surveys and many other available reports, experience with rhesus monkey colonies indicates that these animals are extraordinarily sensitive, cynomolgus monkeys less so. Autopsy reports on the pathology of green monkeys that had been housed with tubercular rhesus monkeys showed that the former had incurred little or no disease (Schmidt, 1972).

The high susceptibility of rhesus monkeys to natural infection with tuberculosis makes them poor risks as laboratory animals. Schmidt (1956) reported that a fulminating widespread incidence in one colony resembled those which have occurred in some human populations. Within fifteen months, the disease had spread from one tuberculin-hypersensitive animal with detectable lesions to all the other monkeys in the cage, and within six months of the first positive tuberculin reaction, 75 percent of the animals were dead. Whereas roentgenographic evidence of pulmonary tuberculosis could not always be obtained at the same time, all of the animals with visible pulmonary lesions died two to four months later.

Not all animals become tubercular soon after natural infection. But knowing that all *can* be infected, Schmidt et al. (1955) performed a series of systematic studies that have since become models of epidemiological experimentation.

The authors first induced pulmonary tuberculosis in rhesus monkeys by intratracheal instillation of as few as 10 to 1,000 tubercular bacilli. Within two to four weeks, all animals had become tuberculin hypersensitive; within four to six weeks, all showed pulmonary infiltrates on roentgenograms; and within six to twelve weeks, all had died. All had developed large areas of caseating pneumonia, which subsequently led to necrosis and cavitation. Regardless of minor modifications which were due to the different strains of tubercular bacilli used, the results were the same. Using these always reproducible methods of infection, Schmidt and his colleagues have studied the various parameters of the biology of tuberculosis as thoroughly as others have done in other diseases.

In young, even immature rhesus monkeys, Schmidt also induced a disease similar to the fulminating infection that afflicts human white infants and black young adults. It is significant that the disease caused in older animals is similar to the chronic cavitary infection found in adult human beings. It should be emphasized here that the characteristic properties of the disease induced by Schmidt (1956) were similar to those of the naturally acquired disease.

Within twenty-eight days after the monkeys had been challenged with approximately 800 viable units, the populations of organisms in their lungs ranged from 400,000 to 20,000,000 viable units, and in their lymph nodes from 20,000 to 10,000,000 viable units (Schmidt, 1972). Doubtless, the increased populations reflect the high susceptibility of these monkeys to the pathogen. In different subjects which had received challenges of 500 viable units, the survival times were as short as seven weeks and as long as twenty-two weeks, and the amount of lung tissue destroyed at death varied considerably. Apparently this variability reflects innate

differences in the genetic capacities of the animals to mobilize their limited immune reactions to fight off the disease.

To find effective ways of controlling the disease, Schmidt et al. (1955) inoculated 120 monkeys intratracheally with minute amounts of tubercle bacilli. Within three weeks, roentgenographic evidence showed that all but one had positive tuberculin hypersensitivity and pulmonary lesions. After the disease level of each animal had been assessed, the monkeys were evenly distributed among seven groups of sixteen or nineteen animals each and treated for six months with streptomycin or isoniazid mixed with their food. The six untreated monkeys in each group died of pulmonary tuberculosis within ninety-six days, but only two of the treated animals died, and these were on low doses of isoniazid. Analyses of the radiologic clearance of the lung lesions, gross examination at autopsy, histologic evaluation, and the recovery of tubercle bacilli from residual lesions confirmed that isoniazid was therapeutically sound but that a number of dosage-dependent variations could be expected. The dosage was later standardized by intramuscular injection or oral intubation, and the results were very constant.

Schmidt (1972) reminds us that the ability of rhesus monkeys to separate "good" from "bad" drugs is enormously important. If used judiciously, some powerful drugs can bring us closer to the development of more effective means of controlling human tuberculosis than are available today. If these studies are sufficiently funded to guarantee their continuation, the devastating world health problem tuberculosis may eventually be solved.

CONCLUDING REMARKS

The subjects treated in these chapters, sometimes briefly, sometimes in depth, are timely and of some importance for three reasons. First, biomedical scientists have at last become fully aware of the importance of using nonhuman primates in research. This awareness, however, is a mixed blessing because many contemporary biological scholars are highly specialized men and women who know much about biochemistry and molecular biology but often practically nothing about such fundamental subjects as natural history, taxonomy, anatomy, and behavior. As a consequence, these specialists sometimes use their primate subjects with a wanton disregard for their immense value and with as little regard for them as living creatures as if they were inert culture dishes. Thus, it is evident that worldwide regulations must be established to control the use of these animals.

Second, regardless of regulations, increasing numbers of scientists will be using primates as experimental models in biomedical research. It is my hope that they would profit by reading even such a selective and brief account as this.

Finally, biomedical research will endure as long as man himself. The cure or even complete eradication of one disease has a way of letting down the barriers to the invasion of others. Thus, if cancer and severe cardiovascular disorders were to be suddenly abolished, man would still be the prey of disease. He might even be subject to maladies as yet unknown, the dubious reward for having attained the eldorado of longevity.

In recent years, we have witnessed the phenomenon of degenerative diseases replacing infections as the major threat to man's survival. Most of these noninfectious diseases have multiple etiologies and manifestations (Portman et al., 1975) and usually develop over lengthy periods of time. These two facts impose very special demands on those of us in biomedical research who must select specific animals as experimental models. Since man's chronic diseases are seldom faithfully duplicated in the average, short-lived laboratory animal, we would do well now, while we still have time on our side, to make long-range plans for the wise use of nonhuman primates in medical research. It is the purpose of this volume to alert as many concerned citizens as possible not only to the nature and importance of these animals but especially to the need for a more judicious and frugal use of them.

About the Author

ABOUT THE AUTHOR

William Montagna became director of the Oregon Regional Primate Research Center, Beaverton, Oregon, in 1963. He is also professor and head, Division of Experimental Biology, and professor of dermatology at the University of Oregon Medical School.

Born in Roccacasale, Italy, he emigrated with his family in October, 1927, and became a naturalized United States citizen. He received an A.B. degree from Bethany College in 1936, and the Ph.D. degree in zoology from Cornell University in 1944. After teaching one year at Cornell University and two years at the New York Downstate Medical School, he joined the faculty of Brown University in 1948, was made professor of biology in 1952, and in 1960 was awarded the chair of L. Herbert Ballou University Professor. In 1950, Dr. Montagna initiated an annual Symposium on the Biology of Skin, which was destined to attract the world leaders in the field. It has just celebrated its silver anniversary.

His principal research interests are in the male reproductive system, the biology of skin, and primatology. He has written

and edited more than twenty-five books, among them a textbook *Comparative Anatomy*, *The Structure and Function of Skin*, now in its third edition, and *Man*, a widely known and adopted book on the biological, sociological, and cultural aspects of man. He has published over 250 original papers on subjects ranging from ornithology to disorders of the skin.

Dr. Montagna has lectured extensively in the United States and abroad and has received many honors and awards. His biographical sketch appears in *Who's Who in the East*, *Who's Who in the West*, *Who's Who in America*, and *Who's Who in the World*. He is also listed in the National Register of Prominent Americans and International Notables, and in 1975 he received the Community Leaders and Noteworthy Americans Bicentennial Memorial Award. He has been president of the Society for Investigative Dermatology. He received the Stephen Rothman Award for distinguished achievement in Investigative Dermatology from that society in 1972, and was presented the Gold Medal Award for Meritorious Achievement in the Biological Sciences from the Università degle Studi, Sassari, Italy.

Dr. Montagna has been decorated three times by the Italian government: the Cavaliere della Repubblica Italiana in 1963; the Cavaliere Ufficiale della Repubblica Italiana in 1969; and in 1975, the highest honor of all, the Commendatore della Repubblica Italiana.

References

REFERENCES

Beach, F. A. (1947). A review of physiological and psychological studies of sexual behavior in mammals. *Physiol. Rev.* 27:240-307.

Beach, F. A. (1950). The snark was a boojum. *Am. Psychol.* 5:115-124.

——— (1969). Locks and beagles. *Am. Psychol.* 24:971-989.

Buettner-Janusch, J. (1966). *Origins of Man.* New York, London, Sydney: John Wiley & Sons.

Bugher, J. C. (1951). The mammalian host in yellow fever. In G. K. Strode (ed.), *Yellow fever.* New York: McGraw-Hill.

Carpenter, C. R. (1964). *Naturalistic behavior of nonhuman primates.* University Park: Pennsylvania State University Press.

Clarke, B. (1975). The causes of biological diversity. *Sci. Am.* 233:50-60.

Davis, N. C. (1931). Transmission of yellow fever; further experiments with monkeys of New World. *Am. J. Trop. Med.* 11:113-125.

Dolinow, P. (1971). The living nonhuman primates. In P. Dolinow and V. M. Sarich (eds.), *Background for man.* Boston: Little, Brown.

Eaton, G. G. (1976). The social order of Japanese macaques. *Sci. Am.* In press.

Eaton, G. G., R. W. Goy, and C. H. Phoenix (1973). Effects of testosterone treatment in adulthood on sexual behaviour of female pseudohermaphrodite rhesus monkeys. *Nature [New Biol.]* 242:119-120.

Eisenberg, J. (1966). The social organization of mammals. *Handbook of Zoology* (Berlin) 10:1-92.

Ford, C. S., and F. A. Beach (1951). *Patterns of sexual behavior.* New York: Ace Books.

Goldfoot, D. A., M. A. Kravetz, R. W. Goy, and S. K. Freeman (1976). Lack of effect of vaginal lavages and aliphatic acids on ejaculatory responses in rhesus monkeys: Behavior and chemical analyses. *Horm. Behav.* 9:1-27.

Gresham, G. A. (1973). The use of primates in cardiovascular research. In G. H. Bourne (ed.), *Nonhuman primates and medical research*. New York and London: Academic Press.

Harding, R. S. O. (1973). Predation by a troop of olive baboons (*Papio anubis*). *Am. J. Phys. Anthropol.* 38:587-591.

Howard, C. F., Jr. (1972a). Spontaneous diabetes in *Macaca nigra*. *Diabetes* 21: 1077-1090.

——— (1972b). Spontaneous diabetes in the colony of Celebes apes. *Primate News*, 10:3-12.

——— (1974a). Diabetes in *Macaca nigra*: Metabolic and histologic changes. *Diabetologia* 10:671-677.

——— (1974b). Spontaneous diabetes in monkeys. *Primate News* 12:3-6.

——— (1975). Diabetes and lipid metabolism in nonhuman primates. *Adv. Lipid Res.* 13:91-134.

Jay, P. C. (1962). Aspects of maternal behavior among Langurs. *Ann. N.Y. Acad. Sci.* 102(2):468-476.

Johnson, D. F., and C. H. Phoenix (1976). The hormonal control of female sexual attractiveness, proceptivity, and receptivity in rhesus monkeys. *J. Comp. Physiol. Psychol.* 90:473-483.

Kenard, M. A., C. R. Schroeder, J. D. Trask, and J. R. Paul (1939). A cutaneous test for tuberculosis in primates. *Science* 89:442-443.

Kirchheimer, W. F., and E. E. Storrs (1971). Attempts to establish the armadillo (*Dasypus novemcinctus*, Linn.) as a model for the study of leprosy. *Int. J. Lepr.* 39:693.

Koch, R. (1898). Berichte über die Ergebnisse der Expedition des Geheimermedicinalrathes Dr. Koch in Schutzgebiete von Deutsch-Ostafrika. *Zentralbl. Bakteriol., Parasitenk. Infektionskr., Abt. 1*, 24:200-204.

——— (1900). Zweiter Bericht über die Thätigkeit der Malaria Expedition. *Deut. Med. Wochenschr.* 26:88-90.

Laveran, A. (1880). Presentation by M. Colin of a manuscript entitled "Note sur un nouveau parasite trouvé dans le sang de plusiers malades atteints de fièvre palustre." *Bull. Acad. Med., Paris* 9:1235-1236.

McGill, H. C., Jr., J. P. Strong, R. L. Holman, and N. T. Werthessen (1960). Arterial lesions in the Kenya baboon. *Circ. Res.* 8:670-679.

McNulty, W. P., Jr., W. C. Lobitz, Jr., F. Hu, C. A. Maruffo, and A. S. Hall (1968). A pox disease in monkeys transmitted to man. *Arch. Dermatol.* 97:286-293.

McNulty, W. P., and M. R. Malinow (1972). The cardiovascular system. In R. N. T-W-Fiennes (ed.), *Pathology of simian primates*, Part I. Basel: S. Karger.

McSherry, C. K., F. Glenn, and N. B. Javitt (1971). Composition of basal and stimulated hepatic bile in baboons, and the formation of cholesterol gallstones. *Proc. Natl. Acad. Sci. U.S.A.* 68:1564-1568.

Malinow, M. R., and C. A. Maruffo (1966). Naturally occurring atherosclerosis in howler monkeys (*Alouatta caraya*). *J. Atheroscler. Res.* 6:368-380.

Malinow, M. R., and C. A. Storvick (1968). Spontaneous coronary lesions in howler monkeys (*Alouatta caraya*). *J. Atheroscler. Res.* 8:421-431.

Medawar, P. B. T. (1973). *The hope of progress: A scientist looks at problems in*

philosophy, literature, and science. Garden City, N.Y.:Anchor Books, Anchor Press/Doubleday.

Michael, R. P., and E. B. Keverne (1968). Pheromones in the sexual status of primates. *Nature* (London) 218:746-749.

———— (1970). A male sex-attractant pheromone in rhesus monkey vagina secretions. *J. Endocrinol.* 46:xx-xxi.

Montagna, W. (1972). The skin of nonhuman primates. *Am. Zool* 12:109-124.

Napier, J. (1962). Monkeys and their habitats. *New Scientist* 15(295):88-92.

Osuga, T., K. Mitamura, S. Miyagawa, N. Sato, S. Kintaka, and O. W. Portman (1974a). A scanning electron microscopic study of gallstone development in man. *Lab. Invest.*, 31:696-704.

Osuga, T., and O. W. Portman (1971). Experimental formation of gallstones in the squirrel monkey. *Proc. Soc. Exp. Biol. Med.* 136:722-726.

———— (1972). Relationships between bile composition and gallstone formation in squirrel monkeys. *Gastroenterology* 63:122-133.

Osuga, T., O. W. Portman, K. Mitamura, and M. Alexander (1974b). A morphologic study of gallstone development in the squirrel monkey. *Lab. Invest.* 30: 486-493.

Phoenix, C. H. (1973a). Sexual behavior of rhesus monkeys after vasectomy. *Science* 179:493-494.

———— (1973b). The role of testosterone in the sexual behavior of laboratory male rhesus. In C. H. Phoenix (ed.), *Primate reproductive behavior*, vol. 2. Symposia of the Fourth International Primatological Congress. Basel: S. Karger.

———— (1974). The effects of dihydrotestosterone on the sexual behavior of castrated male rhesus monkeys. *Physiol. Behav.* 12:1045-1055.

Phoenix, C. H., A. K. Slob, and R. W. Goy (1973). Effects of castration and replacement therapy on the sexual behavior of adult male rhesuses. *J. Comp. Physiol. Psychol.* 84:472-481.

Portman, O. W. (1970). Arterial composition and metabolism: Esterified fatty acids and cholesterol. *Adv. Lipid Res.* 8:41-114.

Portman, O. W., and D. R. Illingworth (1975). Arterial metabolism in primates. In J. P. Strong (ed.), *Primates in medicine*, vol. 9:145-223. *Atherosclerosis in nonhuman primates.* Basel: S. Karger.

Portman, O. W., T. Osuga, and N. Tanaka (1975). Biliary lipids and cholesterol gallstone formation. *Adv. Lipid Res.* 13:135-194.

Premack, A. J., and D. Premack (1972). Teaching language to an ape. *Sci. Am.* 227:92-99.

Reed, W. (1902). Recent researches concerning the etiology, propagation and prevention of yellow fever, by the United States Army Commission. *J. Hyg.* 2: 101.

Resko, J. A. (1972). Sexual behavior and testosterone concentrations in the plasma of the rhesus monkey before and after castration. *Endocrinology* 91:499-503.

Rowell, T. (1972). *The social behavior of monkeys.* Kingsport, Tenn.: Penguin.

Schaller, G. (1963). *The mountain gorilla.* Chicago: University of Chicago Press.

Schmidt, L. H. (1956). Some observations on the utility of simian pulmonary tuberculosis in defining the therapeutic potentialities of isoniazid. *Am. Rev. Tuberc. Pulm. Dis.* 74(Part II):138-159.

——— (1972). Improving existing methods of control of tuberculosis: A prime challenge to the experimentalist. *Am. Rev. Respir. Dis.* 105:190-191.

——— (1973). Infections with *plasmodium falciparum* and *plasmodium vivax* in the owl monkey—model systems for basic biological and chemotherapeutic studies. *Trans. R. Soc. Trop. Med. Hyg.* 67:446-474.

Schmidt, L. H., R. Hoffmann, and P. N. Jolly (1955). Induced pulmonary tuberculosis in the rhesus monkey. Its usefulness in evaluating chemotherapeutic agents. *Trans. 14th VA, Armed Forces Conf. Chemother. Tuberc.* pp. 226-231.

Schultz, A. H. (1961). Some factors influencing the social life of primates in general and of early man in particular. In S. L. Washburn (ed.), *Social life of early man.* Chicago: Aldine.

Simonds, P. E. (1974). *The social primates.* San Francisco and London: Harper and Row.

Small, D. M., and S. Rapo (1970). Source of abnormal bile in patients with cholesterol gallstones. *N. Engl. J. Med.* 283:53-57.

Southwick, C. H., M. R. Siddiqi, and M. F. Siddiqi (1970). Primate populations and biomedical research. *Science* 170:1051-1054.

Storrs, E. E. (1971). The nine-banded armadillo: A model for leprosy and other biomedical research. *Int. J. Lepr.* 39:703.

Strum, S. C. (1975a). Life with the "pumphouse gang." *National Geographic* 47:673-691.

——— (1975b). Primate predation: Interim report on the development of a tradition in a troop of olive baboons. *Science* 187:755-757.

Sturdevant, R. A. L., M. L. Pearce, and S. Dayton (1973). Increased prevalence of cholelithiasis in men ingesting a serum cholesterol-lowering diet. *N. Engl. J. Med.* 288:24-27.

Teleki, G. (1973). The omnivorous chimpanzee. *Sci. Am.* 228:33-42.

van Lawick-Goodall, J. (1971). *In the shadow of man.* Boston: Houghton Mifflin.

Washburn, S. L. (1968). *The study of human evolution.* Condon Lectures. Eugene: Oregon State System of Higher Education.

Washburn, S. L., and I. De Vore (1961). Social behavior of baboons and early man. In S. L. Washburn (ed.), *Social life of early man.* Chicago: Aldine.

Young, M. D., J. A. Porter, Jr., and C. M. Johnson (1966). Plasmodium vivax transmitted from man to monkey. *Science* 153:1006-1007.

Index

INDEX

143

Date Due